1

ZnO Metal-Semiconductor-Metal UV Photodetectors on PPC Plastic with Various Metal Contacts

N.N. Jandow*, H. Abu Hassan, F.K. Yam and K. Ibrahim
Nano-Optoelectronics Research and Technology Laboratory School of Physics Universiti Sains Malaysia, Minden, Penang Malaysia

1. Introduction

The unique and attractive properties of the II-VI compound semiconductors have triggered an enormous incentive among the scientists to explore the possibilities of using them in industrial applications. Zinc Oxide (ZnO) is one of the compound semiconductors of the II–VI family with a direct band gap of 3.37 eV at room temperature, and a large excitation binding energy (60 meV). ZnO is low cost and easy to grow. It is also sensitive to the UV region because of its ultra violet absorbance and has high photoconductivity (Young et al., 2006). These properties make ZnO a promising photonic material for several applications, such as transparent conducting electrodes, surface acoustic wave filters, gas sensors, light emitting diodes, laser diodes and ultraviolet detectors (Kumar et al., 2009; Lim et al., 2006; Janotti et al., 2009; Bang et al.,2003). ZnO is of value and importance due to wide chemistry, piezoelectricity and luminescence at high temperatures. In industry, it can be used in paints, cosmetics, plastics and rubber manufacturing, electronics, and pharmaceuticals [http://www.navbharat.co.in/Clients.htm].

Films of ZnO, indium tin oxide (ITO), and cadmium oxide (CdO) have recently been investigated as transparent conducting oxide (TCO) due to their good electrical and optical properties, their abundance in nature, their optical transmittance (>80%) in the visible region, and they are non-toxic (Angelats, 2006).

ZnO crystallizes in two different crystals structural. The first is the hexagonal wurtzite lattice with lattice constants of a = 3.249 Å and c = 5.207 Å which is mainly used in thin film industry as a transparent conducting oxide (TCO) or as a catalyst in methanol synthesis (Jin, 2003). The second structure, which is well known to geologists, is in the form of rock salt structure which is used in understanding the earth's lower mantle (Hussain, 2008).

ZnO possesses very similar properties to the nitride-based semiconductors such as GaN; yet it has many advantages over GaN (Koch et al., 1985; Paul et al., 2007). For examples: firstly, the growth of high quality single crystal of ZnO is a well understood technology; whereas

---

* Corresponding Author

for GaN it is difficult. Secondly, the ZnO has large exciton binding energy which means that it has a lower threshold power for lasing. Thirdly, it is one of the few oxides that show quantum confinement in an experimentally accessible range of the size of the particles.

2. Deposition of ZnO on various organic substrates

ZnO thin films traditionally have been deposited on mostly inorganic substrates such as quartz, silicon, glass, sapphire, GaAs, fluorite, mica, GaN, Al_2O_3, diamond, NaCl, and InP (Hickernell, 1976; Ianno et al., 1992; Craciun et al., 1994; Jin et al., 2001; Shan et al., 2004; Sans et al., 2004; Ghosh et al., 2004; Zhang et al., 2004; Tsai et al., 2007; Kiriakidis et al., 2007; Wang et al., 2008) by using different techniques (Mahmood et al., 1995; Auret et al., 2007; Rao et al., 2010; Chakraborty et al., 2008; Hwang et al., 2007; Nunes et al 2010; Sofiani et al., 2006).

The preparation of ZnO thin film on flexible substrates such as plastic has so far received much interest due to its wide variety of applications as in flexible sensors and curved detector arrays. Plastic substrates provide lighter, more resistant to damage, flexible, and durable devices (Nandy et al., 2010); these attributes make them suitable for portable devices such as smart cards, personal digital assistants, digital cameras, cell phones, remote control, and circuits camcorders (Brabec et al., 2001). The substrate materials, that are commonly used for the above applications, include polyethylene terephthalate (PET), polyolefin, polytetrafluoroethylene (Teflon), Polycarbonate (PC), polyarylate (PAR), polyestersulfone (PES), polyimide (PI), poly(ethylene naphthalate) (PEN), thermoplastic polymethyl methacrylate (Perpex, Plexiglas), polyethylene naphthalate, and cellulose triacetate (Ma et al., 2008).

Most of ZnO studies have focused on fabrication issues with little attention given to the structural and optical properties of these films on flexible substrates. Interfacial phenomena in ZnO/polymer heterostructures are expected to be very different from those observed for inorganic substrates, there is a need for more comprehensive investigations to assess the potential of ZnO as a material that can be used in flexible electronic applications. Deposition of ZnO on polymers may open a new opportunities for the creation of novel multifunctional polymer/semiconductor heterostructures.

ZnO thin films deposited on plastics had a number of advantages compared with those on inorganic substrates. From the literature, various types of organic substrates have been explored by researchers to deposited ZnO.

Many researchers (Ott & Chang, 1999; Banerjee et al., 2006; Tsai et al., 2006; Lu et al., 2007) have grown ZnO on PET plastic substrates using different techniques for exemple: Ott and Chang (Ott & Chang, 1999) reported that their samples showed highly transparent (T > 80%) and conductive ($\rho \sim 10^{-3}$ Ω cm) and the best film grown on PET had a resistivity of 1.4 ×10^{-3} Ωcm. Banerjee *et al* (Banerjee et al., 2006) deposited ZnO thin films with two different thicknesses (~260 and ~470 nm) for 4 and 5 hours using DC sputtering technique. They found that X-Ray diffraction (XRD) pattern for 4 h deposited ZnO films showed a weak intensity as well as broad peak at (002). But the film deposited for 5 h confirmed the formation of crystalline ZnO and three other peaks originated from (100), (002) and (101) reflections of hexagonal ZnO. Their films showed almost 80% to more than 98% visible transmittance and the bandgap (E_g) values were 3.53 eV and 3.31 eV for the films with (~260 and ~470 nm) thicknesses. Room temperature conductivities of the films were found ranging around 0.05 to 0.25 S cm^{-1} and the maximum carrier concentrations around 2.8×10^{16} and 3.1×10^{20} cm^{-3} with a variation in the deposition time of 4 and 5 h respectively.

On the other hand, Tsai et al (Tsai et al., 2006) reported that their ZnO/PET films showed a strong (002) peak from XRD measurements and the scanning electron microscopy (SEM) morphology of the films showed that no crack or bend appeared on the films. The average transmittance in the visible spectrum was above 80% for both substrates and the value of E_g was found equal to 3.3 eV. The lowest resistivity obtained was 4.0×10^{-4} Ω cm. Lu et al (Lu et al., 2007) found that XRD results of their deposited thin films showed a strong peak at (002) orientation and the root mean square (rms) were in the range 2.63~11.1 nm determined by atomic force microscopy (AFM). The average transmittance of the film was obtained over 80% in the visible spectrum. The lowest resistivity obtained was 4.0×10^{-3} Ωcm.

Whereas, Liu et al (Liu et al., 2007) deposited ZnO thin film at the first time on Teflon substrate by the rf magnetron sputtering. They reported that the XRD pattern of the film showed a strong peak at 2θ = 34.283° which corresponding to the (002) peak with full-width at half-maximum (FWHM) of 0.724° and the film also showed other peaks such as (100), (101), (102), (103) and the crystallite size was equal to 10 nm. The SEM image of ZnO thin film showed that the ZnO film structure consists of some columnar structured grains.

On the other hand, Kim et al (Kim et al., 2009) deposited ZnO films on PC and PES substrates by using rf sputtering system. They studied the effect of sputtering power ranged from 100 to 200 W on the characteristics of the films. XRD patterns of their deposited films showed strong 2θ peaks at 34.4° and they found that the intensities of the ZnO (002) peak and the grain size increased with increasing the sputtering power. The transmittance of the films on both substrates was 80-90%.

Also Liu et al (Liu et al., 2009) prepared Transparent conducting aluminum-doped zinc oxide (ZnO:Al) films on PC substrates by pulsed laser deposition technique at low substrate temperature (room-100 °C. they reported that their experiments were performed at various oxygen pressures (3 pa, 5 pa, and 7 Pa). they studied the influence of the process parameters on the deposited (ZnO:Al) films. X-ray diffraction for their prepared films showed polycrystalline ZnO:Al films having a preferred orientation with the c-axis perpendicular to the substrate were deposited with a strong single violet emission centering about 377–379 nm without any accompanying deep level emission. The average transmittances exceed 85% in the visible spectrum.

Other organic substrates such as polyimide (Paul et al., 2007 and Craciun et al., 1994), polyarylate (Ianno et al., 1992), PEN (Koch et al., 1985) have also been tried by other researchers to deposit ZnO thin film. These ZnO thin films have been used to fabricate thin film transistor (TFT). The fabricated ZnO TFT on various organic substrates showed very encouraging results.

In summary, ZnO thin film deposited on organic substrates by different techniques are in the polycrystalline form with (002) orientation as the dominant peak. The thin films grown on these substrates usually demonstrated good optical transmittance characteristic, i.e. above 80% and the bandgap could be varied from 3.31-3.53 eV. In addition, the carrier concentration is dependent on the growth condition. Overall the ZnO thin films quality on organic substrates is comparable to those in organic substrates.

3. An overview of photodetectors

Photodetectors are basically semiconductor devices that convert the incident optical signal into an electrical signal which is usually revealed as photocurrent. The photodetectors can detect the optical signals over a range of the electromagnetic spectrum that is usually predominantly defined based on the material properties. A detector is selected based on the requirements of a particular application. The general requirements include wavelength of light to be detected, sensitivity level needed, and the response speed. In general, photodetectors respond uniformly within a specific range of the electromagnetic spectrum. Consequently, the wavelength of light detected determines the selection of the photodetector material and the target application and defines the structure (Sze, 2002). This will be explained further in the following sections.

When the incident photons with energy higher than the bandgap energy of a semiconductor; some of them will be absorbed within the semiconductor layer. Such a successful absorption process results in the generation of a free electron-hole pair. The energy gained by the electron which is called the work function must be sufficient to make the electron cross the barrier height between the metal contact and the semiconductor with kinetic energy (E_e). The kinetic energy can be given as (Sze, 2002).

$$E_e = \frac{hc}{\lambda} - \phi_m \tag{1}$$

where c is the light velocity, ϕ_m is the metal work function and λ is the incident wavelength.

Since the photoelectric effect is based on the photon energy hv, the wavelength of interest is related to energy transition ΔE in the device operation, with the following relationship:

$$\lambda = \frac{hc}{\Delta E} = \frac{1.24}{\Delta E(eV)} \qquad (\mu m) \tag{2}$$

where ΔE is the transition of energy levels.

Because the photon energy $hv > \Delta E$ can also cause excitation, Eq. 2 is often the minimum wavelength limit for detection. The transition energy ΔE, in most cases, is the energy gap of the semiconductor. It depends on the type of photodetector, and it can be the barrier height as in a metal-semiconductor (MS) photodiode. Alternatively, the transition energy can be between an impurity level and the band edge as in an extrinsic photoconductor. The type of photodetector and the semiconductor material are normally chosen and optimized for the wavelength of interest. The absorption of light in a semiconductor is indicated by the absorption coefficient. The transition energy does not only determine whether light can be absorbed for photoexcitation, but it also indicates where light is absorbed. A high value of absorption coefficient indicates light is absorbed near the surface where light enters. Whilst, a low value means the absorption is low that light can penetrate deeper into the semiconductor (Sze & Kwok 2006).

3.1 Semiconductors for UV photodetection

UV detection has usually been applied to narrow bandgap semiconductor photodiodes, thermal detectors, photomultiplier tubes (PMT), or charge-coupled devices (CCD) because they exhibit high gain and low noise and they can be rather visible-blind. However, PMT is a fragile and bulky device which requires high power supplies. On the other hand, CCD detectors are slow and their response does not depend on the wavelength. Semiconductor photodetectors require only a mild bias, and can be benefited for their small size and light weight, and being insensitive to magnetic fields. Their low cost, good linearity and sensibility, and capability for high-speed operation make them excellent devices for UV detection (Monroy et al., 2003).

The main disadvantage of these narrow-bandgap semiconductor detectors is device aging due to the exposure to radiation that has much higher energy than the semiconductor bandgap. Moreover, passivation layers, typically SiO_2, reduce the quantum efficiency in the deep-UV range, and are also degraded by UV illumination (Caria et al., 2001)[43]. Another disadvantage of these devices is their sensitivity to low energy radiation, eventually, filters are required to block out visible and infrared photons, resulting in a significant loss of the effective area of the device. Finally, for high-sensitivity applications, the detector active area must be cooled to reduce the dark current. The cooled detector behaves as a cold trap for contaminants which leads to a lower detectivity (Monroy et al., 2003). Such problems make the wide-bandgap semiconductors, such as gallium nitride and ZnO, attractive alternatives. Therefore, ZnO has been used in optoelectronic, high power and high frequency devices.

3.1.1 Wide-bandgap semiconductors for UV photodetection

Detection of ultraviolet (UV) radiation is increasingly becoming important in a number of areas, such as flame detection, water purification, furnace control, UV astronomy, UV radiation dosimetry (Lakhotia et al., 2010).

Even though the responsivity of Si-based optical photodetectors in the UV region is low, they are still being used for light detection (Lakhotia et al., 2010). This has promoted some researchers to use wide direct band gap materials to fabricate optoelectronic devices that are sensitive in the UV region. Hence, GaN-based UV photodetectors have already become commercially available (Hiramatsu et al., 2007). ZnSe-based UV photodetectors, which is another wide direct band gap material, have also been manufactured (Hanzaz et al., 2007). Fabrication and characterization of low-intensity ultraviolet metal-semiconductor-metal (MSM) photodetectors based on AlGaN have also been reported (Gökkavas et al., 2007). ZnO is another semiconductor of wide direct bandgap that is also sensitive in the UV region and is of low cost and easy to manufacture. Therefore, ZnO will be focused in this chapter.

The importance of semiconductor UV PD has expanded the semiconductor industry and emphasized the development of low-light-level imaging systems for military and civilian surveillance applications. These detectors should:

1.   not be sensitive to light at visible wavelengths (commonly referred to as being   solar blind),
2.   have a large response at the wavelength to be detected and have high quantum efficiency,
3.   have a small value for the additional noise introduced by the detector.

There are many different types of semiconductor ultraviolet photodetectors such as: photoconductive or photoconductors detectors, p-n junctions photodiode, and MSM (metal-semiconductor-metal) photodetector (Liu et al., 2010). In the field of optical devices, several trends are pushing research to use new materials. For example, the UV PD is successfully fabricated based on wide bandgap semiconductors (E_g > 3.0 eV). Photon detectors may be further subdivided according to their physical effects that make the detector responsive.

3.1.2 Metal-Semiconductor-Metal (MSM) photodiodes

Metal-semiconductor-metal (MSM) is a type of PD used for UV detection. MSM PDs consist of two interdigitated Schottky contacts which are called fingers deposited on top of an active layer as shown in Figure 1. These devices have a fast response and simple structure compared to other photodetectors of the same active area because of the interdigitated structure which reduces carrier transit time through close spacing of the electrodes, while maintaining a large active area.

The MSM PD operates when the incident light is directed on the semiconductor material between the fingers, electrons will be generated in the conduction band, and thus creating holes in the valance band of the undoped region of the semiconductor. This results in creating a photocurrent by means of one of two processes with the operative process determined by the magnitude of the incident photon energy ($h\upsilon$) relative to the energy bandgap (E_g) of the semiconductor and the metal work function (ϕ_{Bn}). If E_g is greater than $h\upsilon$ and $h\upsilon$ is greater than ϕ_{Bn}, then photoelectric emission of electrons from the metal to the semiconductor occurs. Alternatively, for the second process, if $h\upsilon$ is greater than E_g, then photoconductive electron-hole pairs are produced in the semiconductor. The generated electrons and holes are separated by an electric field intrinsically formed between the fingers.

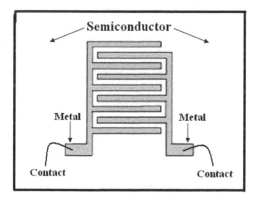

Fig. 1. Top view of an MSM PD planar interdigitated structure

4. ZnO as a UV detector

As mentioned before, ZnO is one of the most prominent semiconductors in the metal-oxide family because of its excellent properties which attract many researcher groups to use this material in UV detection applications. Below, a briefly discussion of ZnO as a UV detector will present and some earlier studies about the fabrication of ZnO as UV detector will also be provided.

ZnO was one of the first semiconductors to be prepared in a rather pure form after silicon and germanium. It was extensively characterized as early as in the 1950`s and 1960`s due to its promising properties (Plaza et al., 2008). Wide band gap semiconductors have so far gained much attention for the last decade because they can be used in optoelectronic devices in the short wavelength and also in the UV region of the electromagnetic spectrum. These

Wide band gap semiconductor	Crystal structure	Lattice parameter (Å)		E_g (eV) at RT	Melting temp. (K)	Excitation binding energy (meV)
		a	c			
ZnO	Wurtzite	3.25	5.206	3.37	2248	60
GaN	Wurtzite	3.189	5.185	3.4	1973	21
ZnSe	Zinc-blende	5.667	-	2.7	1790	20
ZnS	Wurtzite	3.824	6.261	3.7	2103	36
4H-SiC	Wurtzite	3.073	10.053	3.26	2070	35

Table 1. Comparison of different semiconductors (Tüzemen & Gür, 2007; Nause & Nemeth, 2005) Note: Where RT is room temperature, meV is millielectron volt.

Sab.	D.M.	PD T	R. C.	O. C.	B.V. (V)	I_d (A)	I_{ph} (A)	λ (nm)	R (A/W)	Ref
Sapp.	MOCVD	MSM UV	-	Al	5	1×10^{-6}	-	-	1.5	(Liang et al., 2001)
GaAs	rf	p-n I IJ	-	-	~-3.0	-	$~-2 \times 10^3$	325	-	(Moon et al., 2005)
p-Si (100)	rf		Au-Al	In	30	-	-	310	0.5	(Jeong et al., 2004)
Sapp.	Sol–gel	MSM UV	Au	-	-	-	-	350	0	(Basak et al., 2003)
Sapp.	MOCVD	MSM	Al	Al	5	450×10^{-6}	-	-	400	(Liu et al., 2000)
Quar.	rf	MSM UV	Au	Au	3	250×10^{-6}	-	360	30	(Liu et al., 2007)
Sapp.	PA MBE	UV		Al-Ti	20		-	374	1.7	(Mandalapu et al., 2007)
SiO₂	rf	MSM UV	Au	Au	3	1×10^{-3}	-	-	0.3	(Jiang et al., 2008)
Sapp.	rf	MSM UV	Ir	Ir	-	-	-	370	0.2	(Young[a] et al 2007)
Sapp.	rf	MSM UV	Pd	Pd	1	-	-	370	0.1	(Young[b] et al., 2007)
Sapp.	MBE	MSM UV	Ru	Ru	-	8×10^{-8}	1.8×10^{-5}	-	-	(Lin et al., 2005)

Sab = Substrate; D.M. = Deposition method; PD T. = PD type; B.V. = Bias voltage R C. = Rectifying contacts; O. C. = Ohmic contact; I_d = Dark current; I_{ph}= Photocurrent; λ = Wavelength; Sapp. = Sapphire; Quar. = Quartz; ED = Electrochemical deposition; PAMBE = Plasma-assisted molecular-beam epitaxy; p-nHJ = p-n homojunction

Table 2. Summarizes the characteristics of ZnO UV PDS reported by other researchers fabricated by different methods

semiconductors which include ZnO, GaN, ZnSe, ZnS, and 4H-SiC have shown similar properties with their crystal structures and band gaps (Tüzemen & Gür, 2007).

Table 1 below shows a summary of some of the important properties of these wide band gap semiconductors. Initially, ZnSe and GaN based technologies made significant progress in the blue and UV light emitting diode and injection laser. No doubt, GaN is considered to be the best candidate for the fabrication of optoelectronic devices. However, ZnO has great advantages for light emitting diodes (LEDs) and laser diodes (LDs) over the currently used semiconductors. Recently, it has been suggested that ZnO is promising for various technological applications, especially for optoelectronic short wavelength light emitting devices due to its wide and direct band (Nause & Nemeth, 2005).

From the literature ZnO UV PDs have been widely investigated by many researchers on different substrates through different methods. Table 2 compiles and summarizes the structural of ZnO UV PDS reported by other researchers.

5. Schottky barrier height calculation of MSM PD

A MSM PD is a unipolar device with two back-to-back Schottky junctions formed on the same semiconductor surface. Under the application of bias voltage, one of the diodes becomes reverse biased, forming a depletion region that tends to sweep out photocarriers. The other diode becomes forward biased, allowing the collected photocurrent to flow out just as an ohmic contact. Under sufficiently high bias voltage, the depletion region extends and touches the small space-charge region under the forward biased electrode.

Figure 2 shows the energy-band diagram of the MSM PD in the biased state. The vertical displacement of the electrode metals indicates the bias voltage applied to the device resulting in the forward biased condition of the left-hand metal-semiconductor interface and the reverse bias of the right-hand interface. Upon biasing, the semiconductor between the electrodes becomes fully depleted of free carriers. The reversed biased interface prevents the current from flowing through the device when there is no optical signal. The depleted regions between the electrodes are the photodetector active regions. As mentioned before, a photon with energy greater than the band gap of the semiconductor will be absorbed by an electron; this electron will get excited to the conduction band as shown by the process labeled as (1) in Figure 2. The photogenerated electron and hole are swept by the high applied fields to the positive and negative electrodes resulting in an electronic output signal (Haas, 1997).

One can assume that other current transport processes also contribute to the movement of electrons within the depletion region and across the barrier in Schottky contacts to determine the barrier height. The equations usually used to determine the barrier height in a Schottky diode. Assuming pure thermionic emission and V>3KT, the general I–V equations usually used to determine the barrier height in a Schottky diode are represented by (Daraee et al., 2008)

$$I = I_0 \exp[qV / (nKT)] \tag{3}$$

$$I_0 = A^* A T^2 \exp[-q\phi_B / (KT)] \tag{4}$$

where I_0 is the saturation current, n is the ideal factor, K is the Boltzmann's constant, T is the absolute temperature, ϕ_B is the barrier height, A is the area of the Schottky and A^* is the effective Richardson coefficient. Eq. 4 indicates the dependence of the barrier height on the saturation current (I_0). The theoretical value of A^* can be calculated using Eq. (5) below.

$$A^* = 4\pi m^* qK^2 / h^3 \tag{5}$$

where h is Planck's constant and $m^* \sim 0.27m_o$ is the effective electron mass for n-type ZnO so that $A^* \sim 32\ A/cm^2K^2$ (Liang et al 2001).

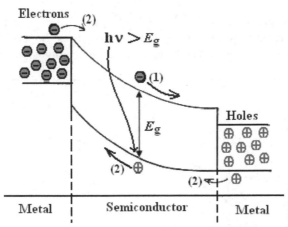

Fig. 2. Energy-band diagram of MSM detector indicating; (1) the photogeneration of signal charges and (2) the thermally-generated carriers overcoming barrier heights adding to device dark current

MSM-PD performance is critically dependent on the quality of the Schottky contacts. Therefore, it is necessary to measure the Schottky barrier height of the actual contact which is a constituent part of the MSM PD under investigation. As mentioned before, a MSM-PD essentially consists of two Schottky contacts connected back-to-back. When a bias is applied within the MSM; this will put one Schottky barrier in forward direction (anode) and the other is reverse direction (cathode).

Following the analyses of Sze (Sze et al., 1971), one can get the approximate dark current formula of the MSM PD. The current transported over the Schottky barrier height as a function of the applied voltage by considering both electron and hole current components here has the general expression as (Sze & Kwok 2006)

$$I_{da} = A_1 A_n^* T^2 \exp\left(\frac{-q\phi_{Bn}}{KT}\right) + A_2 A_p^* T^2 \exp\left(\frac{-q\phi_{Bp}}{KT}\right) \tag{6}$$

where A_1 and A_2 are the anode and cathode contact areas respectively; A_n^* and A_p^* are the effective Richardson constants; and ϕ_{Bn} and ϕ_{Bp} are the barrier heights for electrons and holes respectively. For a semiconductor with wide bandwidth, the Schottky barrier of the hole is high, and the hole is a minor carrier. Hence the hole current can be neglected, assuming that the dark current can mostly consist of the electron current (Jun at al., 2003). The dark current of the MSM PD is approximately as shown below (Yam & Hassan, 2008)

$$I = I_0 \exp[qV / (nKT)][1 - \exp(-qV / KT)] \tag{7}$$

Thus, Equation (6) can be re-written follows:

$$\frac{I \exp[qV / (KT)]}{\exp[qV / (kT)] - 1} = I_0 \exp[qV / (nKT)] \tag{8}$$

Based on Eq. (8), the plot of $\ln\{I \exp(qV / \{KT\}) / [\exp(qV / \{KT\}) - 1]\}$ vs. V results in a straight line; I_0 is derived from the interception with y-axis as shown in Figure 3. Schottky barrier height ϕ_B at the MS interface can be obtained by substituting I_0 value in Eq. (4).

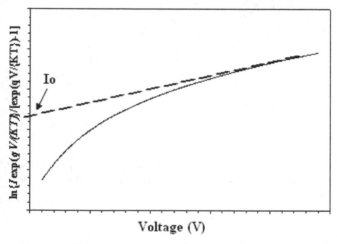

Fig. 3. $\ln\{I \exp(qV / \{KT\}) / [\exp(qV / \{KT\}) - 1]\}$ vs.V of a MSM PD

6. Photodetectors characteristics

There are many characteristics that describe the performance of a PD. These performance characteristics indicate how a detector responds. The response of the detector should be great at the wavelength to be detected. Whilst, the additional noise created by the detector should be small. The response speed should be high so that the variations in the input optical signal can also be detected. The performance characteristics of the photodetectors are summarized in the following subsections.

6.1 Responsivity (R)

Responsivity is defined as the ratio of the photocurrent output (in amperes) to the incident optical power (in watts). It is expressed as the absolute responsivity in amps per watt (A/W). Therefore, responsivity is to measure the effectiveness of the detector for converting the electromagnetic radiation to the electrical current. Responsivity depends on wavelength, bias voltage, and temperature. Reflection and absorption characteristics of the detector's material change with wavelength and hence the responsivity also changes with wavelength (Sze & Kwok 2006).

The responsivity in terms of the photo current can be written as:

$$R = \frac{I_{ph}}{P_{inc}} = \frac{\eta q}{h\nu} \tag{9}$$

where I_{ph} is the photo current (A), P_{inc} is the incident optical power (W), η is the external quantum efficiency of the PD.

The responsivity in terms of wavelength can also be written as:

$$R = \frac{\eta \lambda q}{hc} = \frac{\eta \lambda (\mu m)}{1.24} \ (A/W) \tag{10}$$

where λ is the incident wavelength.

Responsivity is an important parameter that is usually specified by the manufacturer. Through responsivity, the manufacturer can determine how much detector's output is required for a specific application.

6.2 Quantum efficiency

Quantum efficiency of a photodetector (QE) (η %) is defined as the ratio of the electron generation rate to the photon incidence rate. QE is related to the photodetector responsivity by the following equation (Sze & Kwok 2006):

$$\eta = \frac{I_{ph}/q}{P_{inc}/h\nu} = \frac{I_{ph}}{q} \cdot \frac{h\nu}{P_{inc}} \tag{11}$$

7. Characteristics of ZnO thin films deposited on PPC plastic

From the literature, the deposition of ZnO thin film on organic substrate such as PPC has not been reported by other groups. Therefore, PPC shows great potential as a substrate for ZnO thin film. The PPC has been chosen for its excellent properties such as low cost, high dielectric strength and high surface resistivity; and it is considered as one of the important polymers for biomedical and engineering application (Zhang et al., 2007). It also exhibits high transparency and superior mechanical strength (Song et al., 2009). These attractive

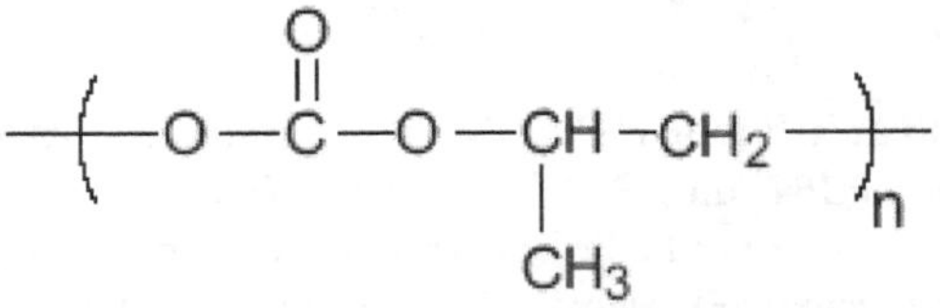

Fig. 4. The chemical structure of Poly (propylene carbonate) (PPC)

properties make the PPC a good substrate for a variety of microelectronic applications. Figure 4 shows the chemical structure of PPC plastic (Lu et al., 2005). This section briefly presents the structural, optical and electrical properties of ZnO deposited on PPC substrate.

7.1 The structural properties

ZnO thin film had been deposited on PPC plastic substrate with thickness of the 1μm. In order to study the structure of the film, XRD diffraction pattern of the deposited ZnO thin film on PPC plastic substrate was measured as shown in Figure 5. The XRD pattern shows that the film is ZnO as compared to the International Center for Diffraction Data (ICDD) library, and it is oriented in c-axis, which is the preferred orientation axis for such a material.

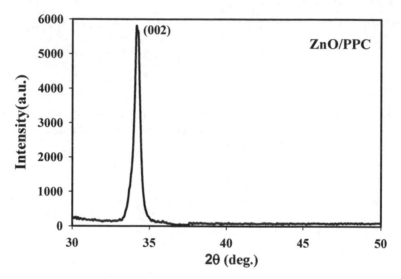

Fig. 5. XRD diffraction for ZnO thin film on PPC plastic substrate

The film shows a strong peak at 2θ = 34.27º which is correlated to the characteristic peak of the hexagonal ZnO (002) with full width at half maximum (FWHM) of 0.31º. The small FWHM of the ZnO (002) XRD peak again indicates good crystal quality of the sample (Jandow et al., 2010a).

The grain size also calculated from the FWHM of the XRD spectrum based on Scherrer formula as in Eq. 12 (Tan et al., 2005) was found to be about 26.8 nm.

$$D = \frac{K\lambda}{B\cos\theta}$$ (12)

where B is the full width at half maximum[(FWHM) in radians] intensity of XRD, λ is the X-ray wavelength (Cu $K\alpha$ =0.154 nm), θ is the Bragg diffraction angle, and K is a correction factor which is taken as 0.9 (Tan et al., 2005). Thus we can conclude that the film deposited on PPC plastic is nanostructured crystal. This finding is in agreement with the result reported by Myoung *et al* (Myoung et al., 2002).

The surface morphology of the film was studied using SEM as shown in Figure 6. The figure shows that the ZnO has smooth surface morphology and relatively smaller particles, which are well connected to each other; it strongly adheres to the substrate and has tightly bounded particles. Inset in Figure 6 shows the SEM image reported by Ergin *et al* (Ergin et al., 2009). A close visual inspection reveals that the prepared sample in this work has similar surface morphology as reported by Ergin. These good surface properties have strong effect on the optical properties such as transmittance and absorbance of the UV light when this material is used as UV detector (Jandow et al., 2010b).

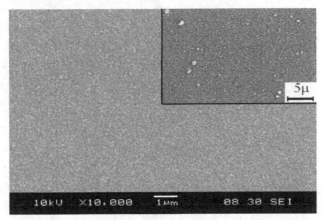

Fig. 6. SEM image of ZnO film on PPC substrate

The elemental analysis of the sample was investigated by EDX as shown in Figure 7.The EDX result shows that Zn, O and C elements were present in the sample. Zn and O elements came from ZnO film, on the other hand, C element was not anticipated in the film; obviously, the presence of C was due to the PPC substrate. Similarly, Ergin *et al* (Ergin et al., 2009) who studied the properties of ZnO deposited on glass substrate pointed out that Si and Ca elements were not expected to be in ZnO film and these two elements could have come from the glass substrates. The inset in Figure 7 shows the EDX image reported by them.

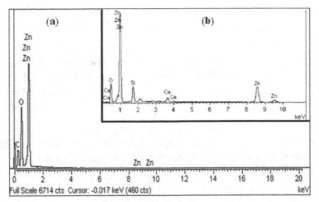

Fig. 7. (a) EDX image of ZnO film on PPC substrate. (b) Inset shows the EDX image from Ref. (Ergin et al., 2009)

The film surface was examined by AFM as shown in Figure 8. A typical AFM image shows that ZnO thin film consists of some columnar structure grains with root mean square (rms) equal to 10 nm; this nanostructure of the film surface may be useful in the absorption of the UV light when this material is used as a UV detector (Jandow et al 2010c).

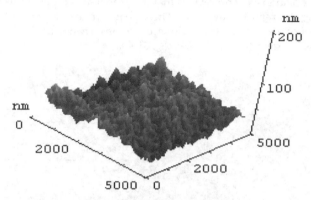

Fig. 8. AFM image of ZnO on PPC substrate (rms =10 nm)

7.2 The optical properties

The PL of the prepared ZnO thin film on PPC plastic was measured at room temperature. The result is given in Figure 9. Two luminescence peaks can be found in the figure. The first luminescence peak is the UV emission of ZnO thin films at 379.5 nm and as mentioned before it corresponds to the near band edge emission (NBE) due to the electronic transition from the near conduction band to the valence band as reported by Young *et al* and Gao and Li (Young et al., 2006; Gao &Li 2004).

The other luminescence peak is the blue-green emission ranged from 452.0 to 510.0 nm as shown in the inset in Figure 9, which is due to the defect related to deep level emission (Tneh et al., 2010). This result is in agreement with Wei *et al.* and Wu *et al* (Wei et al., 2007; Wu et al., 2007), which is attributed to the transition of electron from defect level of Zn interstitial atoms to top level of the valence band. In addition, the high UV to the visible emission ratio indicates a good crystal quality of the film which means a low density of surface defects (Jandow et al., 2010b).

The transmission and absorption spectra of ZnO film are shown in Figure 10. It can be seen that the transmission values of the film are low at short wavelengths ($\leq$ 380nm) and high at long wavelengths. Therefore, the film behaved as an opaque material because of its high absorbing properties at short wavelengths as shown in the same figure and as a transparent material at long wavelengths.

This situation is related to the energy of the incident light; when energies of photons are smaller than the bandgap of ZnO film, they are insufficient to excite electrons from the valence band to the conduction band. However, ZnO has oxygen vacancies and interstitial Zn atoms, which act as donor impurities (Jandow et al., 2010d).

These impurities may be ionized by these low energies, so the film has low absorbance and high transmittance values at long wavelengths. The transmittance values increased higher than 80% in the visible region remarkably as the wavelength increased and this range refers to the fundamental absorption region (Shan et al., 2005). This indicates that the film could be used as transparent windows for UV light or as electrodes in a metal-semiconductor-metal MSM PDs (Jandow et al., 2010e).

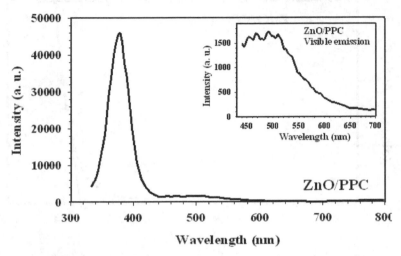

Fig. 9. Room temperature PL spectrum of DC- sputtered ZnO on PPC substrate. Inset shows the expansion of visible emission of ZnO thin film

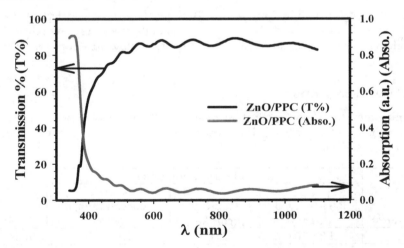

Fig. 10. Transmission and absorption spectra of ZnO thin film on PPC substrate

As mentioned earlier; ZnO is a wurtzite structure semiconductor with a direct band gap of 3.37 eV at room temperature. The absorption coefficient of the direct band gap material is given by the Eq. (Shan et al., 2005)

$$\alpha(h\nu) \propto (h\nu - E_g)^{1/2} \tag{13}$$

The dependence of $(\alpha h\nu)^2$ against the photon energy $(h\nu)$ is plotted as in Figure 11. By extrapolating the linear part of the plot to $(\alpha h\nu)^2 = 0$ the value of the energy gap was found to be about 3.33 eV.

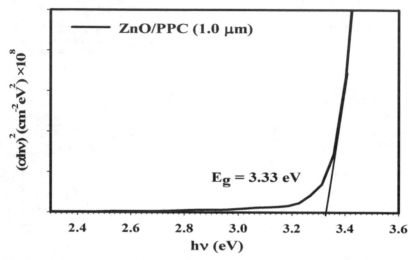

Fig. 11. The bandgap derivation for ZnO thin film on PPC substrate from the dependences of $(ah\nu)^2$ on $h\nu$

Similar results were reported by Banerjee *et al* (Banerjee et al., 2006) who deposited ZnO thin film on PET plastic substrate and also by many other authors who prepared their ZnO thin films on glass and silicon (Gümüş et al., 2006; Khoury et al., 2010; Kang et al., 2007; Lai et al., 2008).

7.3 The electrical properties

The Hall measurements show that the film is n-type with resistivity, ρ, of about 1.39×10^{-1} Ω-cm and mobility, μ, of 26 cm^2/V-s. The carrier concentration, n, was measured and it was found to be 1.72×10^{18} cm^{-3}.

	ZnO Electrical properties			
Substrate	Resistivity, ρ (Ω-cm)	Mobility, μ(cm^2/V-s)	Carrier concentration, n (cm^{-3})	Ref.
PPC	1.39×10^{-1}	26.00	1.72×10^{18}	In this work
PET	1.00×10^{-3}	N.A.	N.A.	(Ott et al., 1999)
PET	N.A.	19.82	2.80×10^{16}	(Banerjee et al., 2006)
PET	4.0×10^{-3}	N.A.	N.A.	(Tsai et al., 2006)

Table 3. Some of the reported ZnO electrical properties on organic substrates

From the literature, the electrical properties of ZnO thin film deposited on organic substrates have rarely been reported. Table 3 summarizes some of the reported electrical properties. From the table, it was found that the resistivity value for our sample is about two orders of magnitudes higher than typical reported values (Ott et al., 1999; Tsai et al., 2006) and the mobility, μ, is 1.3 times higher than the results in (Banerjee et al., 2006). On the other hand, the carrier concentration, n, of our sample is about two orders of magnitudes higher as compared to the reported value in Ref (Banerjee et al., 2006).

8. The characteristics of ZnO UV PDs prepared on PPC

ZnO UV PDs with different metal contacts i.e. Pd, Ni and Pt have been fabricated on PPC plastic substrates. Figure 12 shows the fabricated ZnO UV PD with Pd contacts. The UV detector fabricated on PPC is very flexible and low cost.

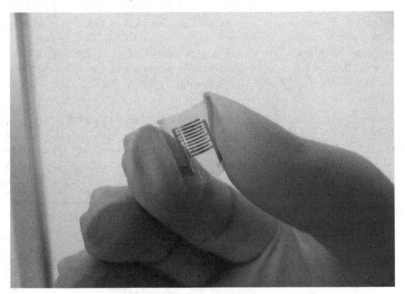

Fig. 12. The ZnO UV PD fabricated on PPC plastic with Pd contacts

8.1 I–V characteristics

Figure 13 shows the I–V characteristics of the fabricated ZnO MSM PDs (Pd/ZnO, Ni/ZnO and Pt/ZnO) on the PPC plastic measured without and with UV illumination (385 nm with power of 58.4 μW). Under dark environment, the current (I_d) at 0.8 volt was equal to 0.44, 1.72 and 1.90 $\square$A and the ideality factor (n) was derived and found to be 1.37, 1.76 and 1.78 for PD with Pd, Ni and Pt contacts respectively..

On the other hand, when the sample was illuminated with UV wavelength of 385 nm with power of 58.4 μW; the photocurrent (I_{ph}) (at same voltage) biased at 0.8V was 5.27, 7.44 and 8.80 μA, and n was found to be 1.21, 1.46 and 1.50, respectively for the fabricated ZnO PD with Pd, Ni and Pt contact electrodes. Table 4 summarized the I-V characteristics with different values of I_d, I_{ph} and n for the PDs with Pd, Ni and Pt metal contacts.

Metal contact	Dark Environment			Illuminated Environment			Change of SBH ϕ_B (meV)
	Dark current, I_d (μA)	Ideality factor, n	SBH, ϕ_B (eV)	Photo current, I_{ph} (μA)	Ideality factor, n	SBH, ϕ_B (eV)	
Pd	0.44	1.37	0.738	5.27	1.21	0.700	38
Ni	1.72	1.76	0.705	7.44	1.46	0.672	33
Pt	1.90	1.78	0.700	8.80	1.50	0.668	32

Table 4. A summary of the I_d, I_{ph} and n for the PDs with Pd, Ni and Pt metal contacts

The results obtained for this study can be explained as follows: when light impinges onto the MSM UV detector, high-energy photons will be absorbed by the ZnO thin film, and with an appropriate bias, photon-generated carriers will drift toward the contact electrodes and a photocurrent will be observed (Young et al., 2006).

Further application of reverse bias acts to increase the electric field magnitude within the depletion region and to reduce the barrier height (Sze & Kwok 2006). When the energy of the incident photon is higher than the bandgap of ZnO, electron-hole pairs are generated inside ZnO thin film by light absorption. At the same time, the electron–hole pairs are separated by the electric field inside the depletion region of the ZnO thin film to generate the photocurrent.

In this study, the application of a bias on each of the Pd, Ni or Pt metallic fingers will create an electric field within the underlying ZnO thin film that sweeps the photo generated carriers out of the depletion region. The speed and the collection efficiency of the device vary, depending upon the magnitude of the applied bias, the finger separation and the average depth at which the photo generated carriers are produced (Jandow et al., 2010b).

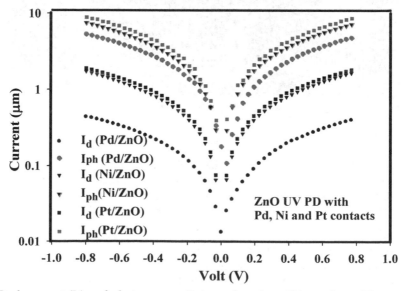

Fig. 13. Dark current (I_d) and photocurrent (I_{ph}) as a function of bias voltage (V) characteristics of the fabricated ZnO MSM PD with different electrodes type

Comparing among the three PDs, PD with Pd has the lowest dark current, this follows by Ni and Pt. the dark current for PT and Ni contacts are about 4.3 and 3.9 times of Pd. The difference of the dark current for PT and Ni contacts is relatively insignificant. This could be anticipated from the SBHs, as shown in Table 4. The difference of SBH value for these contacts is very small i.e. 5 meV.

8.2 Schottky barrier height calculation

As mentioned earlier, one of the most interesting properties of a metal-semiconductor (MS) interface is its Schottky barrier height (SBH), which is a measure of the mismatch of the energy levels for majority carriers across the MS interface. The SBH controls the electronic transport across MS interfaces and is, of vital importance to the successful operation of any semiconductor device (Tung et al., 2001). SBH plays an important role to modulate the dark and photo currents.

In particular, the forward-bias portion of the I-V characteristics has often been used to deduce the magnitude of SBH (Cho et al., 2000). The transport of carriers across a MS interface is very sensitively dependent on the magnitude of the energy barrier, SBH.

SBH for the three PDs could be determined by using Eqs. (3-8). Based on Eq. 8, the plot of $\ln\{I\exp(qV/\{KT\})/[\exp(qV/\{KT\})-1]\}$ vs. V for Pd contact is shown in Figure 14. I_0 was derived from the intercept with y-axis and by substituting this I_0 value in Eq. 4, ϕ_B was found to be 0.738 eV. Similarly for Ni/ZnO and Pd/ZnO, the ϕ_B determined from the plot under dark and illuminated conditions are summarized in Table 4.

Many metals such as Ag, Au, Pd and Pt have been used as Schottky contacts for ZnO, and resulted in SBH of between 0.6-0.8 eV (Gür et al., 2007).

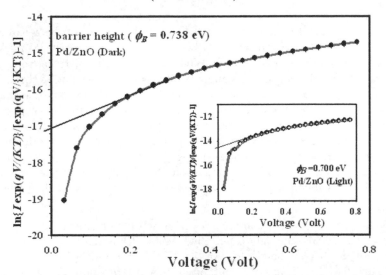

Fig. 14. $\ln\{I\exp(qV/\{kT\})/[\exp(qV/\{kT\})-1]\}$ vs V (under dark condition) of the fabricated ZnO MSM PD with Pd electrodes on PPC substrate. The inset shows ϕ_B calculation under illumination

From Figure 13 it can be found that the ZnO MSM UV PD with Pt electrode has the highest light current. That is because it has the lowest Schottky barrier height at the Pt/ZnO interface. While still the light current for the PD with Pd contact is the lowest for the same reason.

One can find that the difference among the electrical characteristics of the three PDs may mean that their barriers are different. Since the barrier height of the PD with Pd contact is higher than the barrier height for the PDs with Ni and Pt contacts, the integrated number of carriers above the barrier height for the PD with Pd contact will be less than the others, which caused the total current over the barrier to be lower. Eventually the lowest dark current was obtained with the Pd contact compares with the PDs with Ni and Pt contacts.

From Table 4, it was found that the calculated SBH value for the PD with Pd and Ni contacts is higher than that with Pt contact although the work function of Pt is the highest. This indicated that the barrier height is independent of the work function and many research groups (Kim et al., 2010; Wright et al., 2007; Ip et al., 2005; Liu et al 2004; Dong and Brillson, 2008; Kahng, 1063; Vanlaar and Scheer, 1965; Andrews and Phillips, 1975; Mead and Spitzer, 1963; Yıldırım et al., 2010; Roccaforte et al., 2010; Brillson et al., 2008; Rabadanov et al., 1982; Mead, 1966; Coppa et al., 2005) have reported that the barrier heights did not correlate with the metal work functions, suggesting many possible influences such as surface states, surface morphology, and surface contamination play important roles in the electrical properties of the contacts.

The increase of current when PD is illuminated by an UV source can be explained by work function of metal and semiconductor in the energy band diagram (Brillson et al., 2008; Rabadanov et al., 1982; Kim et al., 2010). In this case the metal and semiconductor are taken as Pd and ZnO.

ZnO which has a work function of ϕ_{ZnO}= 4.1 eV is also known to be a natural n-type semiconductor due to the oxygen vacancy which acts as trap center. The Pd work function (ϕ_{Pd} = 5.6 eV) is higher than that of ZnO. When a contact is formed electrons flow from ZnO to Pd metal until the Fermi levels align resulting in band bending as shown in Figure 15, where E_0, E_C, E_{FS}, E_V and $\square$ are the vacuum level, the conduction band, the Fermi level, the valence band and the work function of the semiconductor (S) which in our case is ZnO, respectively, and E_{FM} is the Fermi level of the metal contact (M) which is Pd as shown in the figure, respectively. In ZnO, the trapping mechanism administrates the photoconduction. The oxygen molecules in the ZnO thin film surface capture electrons from the semiconductor. Under UV illumination, the photon energy releases the trapped electrons and also generates photo-induced electrons, causing an enhancement of the current.

The results obtained in this work are in agreement with the results reported by Young *et al* (Young et al., 2006) who fabricated ZnO MSM PD with Ag, Pd and Ni contact electrodes, and the barrier height for Ag/ZnO, Pd/ZnO and Ni/ZnO interfaces were 0.736, 0.701 and 0.613 eV, respectively although the work function for Ag =4.74, Pd =5.60 and Ni =5.42 eV. This result showed that although the work function of Ag is the lowest but the barrier height for Ag/ZnO was the highest, which indicated that the barrier height independent on the work function. Furthermore, Polyakov *et al* (Polyakov et al., 2003) results also showed that Schottky barrier heights were equal to 0.65-0.70 eV from capacitance-voltage measurements by using Au and Ag Schottky contacts on the bulk n-ZnO crystals; Au/ZnO SBH was lower than Ag/ZnO even though the work function of Au 5.47 is higher than the Ag. As well as, Neville and Mead (Neville & Mead 1970) results showed the barrier energy for Au is 0.66 eV and for Pd is 0.60 eV dependent on the forward current-voltage characteristics and as mentioned earlier the work function of Pd is higher than Au.

Before contact   **After contact**

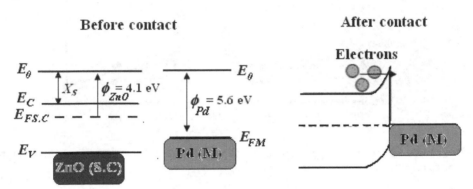

Fig. 15. The energy band diagram of ZnO and Pd.

The independence of SBH on the metal work function has been explained by many researchers (Brillson et al., 2008; Rabadanov et al., 1981; Mead, 1966; Coppa et al., 2005), they pointed out SBH could be affected by the interface structure and the associated interface states

8.3 Responsivity and quantum efficiency

Figure 16 shows the responsivity as a function of the incident wavelength for the three ZnO MSM UV PDs with Pd, Ni and Pt contact electrodes. At 0.8 V and by applying Eq. 9, it was found that the maximum responsivity for the three PDs were found to be 0.08, 0.09 and 0.11 A/W, which corresponds to quantum efficiency (η %) of 27.1 % 31.7 % and 37.7% respectively which are shown in Figure 17. It is observed that the PD with Pt contacts has higher responsivity and quantum efficiency comparing to the PDs with Pd and Ni contacts due to the lowest barrier height and highest photocurrent. Table 5 summarizes the responsivity and quantum efficiency for Pd, Ni and Pt contacts, respectively.

As shown in the figure, the PD responsivity was nearly constant in the UV region ~ 320–360 nm; it started to increase from 365 nm, of which the responsivity peaked at 385 nm and began to decrease whenever became close to the visible region. This can be explained as follows; when ZnO detector is irradiated by UV light with energy higher than the bandgap (3.37 eV for ZnO), electron–hole pairs will be generated, as a result these excess charge carriers contribute to photo current and result in the response to the UV light. The cut-off at wavelength of 385 nm is nearly close to the ZnO energy bandgap of 3.37 eV; the responsivity decreased at the shorter wavelength range due to the decrease of the penetrating depth of the light, resulting in an increase of the surface recombination (Young et al 2007; Jandow et al., 2010a; Yan et al., 2004).

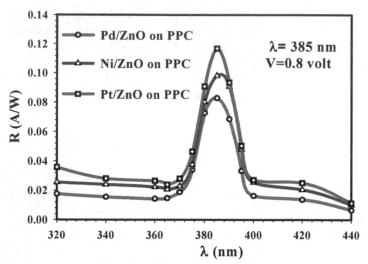

Fig. 16. Measured spectral responsivity of the three fabricated ZnO MSM PDs with different electrodes on PPC substrate

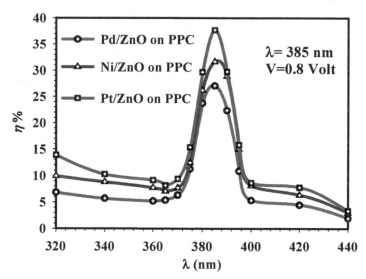

Fig. 17. Quantum efficiency of the three fabricated ZnO MSM PDs with different electrodes on PPC substrate

Metal contacts	Responsivity (R) (A/W)	Quantum efficiency (η %)
Pd	0.082	27.1
Ni	0.098	31.7
Pt	0.116	37.7

Table 5. A summary of the responsivity (R) and quantum efficiency (η %) values for the PDs with Pd, Ni and Pt metal contacts at 0.8 volt

9. Conclusion

In summary, an overview of the characteristics of the deposited ZnO thin film on various organic substrates such as polyethylene terephthalate (PET), polyolefin, polytetrafluoroethylene (Teflon) and Polycarbonate (PC) and their potential applications in various areas has been presented. A review of semiconductor PDs, types of PDs as well as their characteristics has been demonstrated. Apart from that, the properties of the ZnO thin films deposited on PPC plastic substrate have been expressed. ZnO UV detectors prepared on PPC substrate with different electrodes i.e. Pd, Ni and Pt have been fabricated and investigated. The results showed that the deposited ZnO thin film had good structural and optical properties. In addition ZnO UV PDs fabricated on PPC with different metal contacts showed that Pt/ZnO MSM UV PD has the highest quantum efficiency.

10. Acknowledgments

This work was conducted under the grant no. (100/PFIZIK/8/4010) support from Universiti Sains Malaysia is gratefully acknowledged.

11. References

Andrews. J. & Phillips, J. (1975). Chemical bonding and structure of meta semiconductor interfaces. *Phys. Rev. Lett.*, Vol.35, No.1, pp. 56-59

Angelats, L. (2006). "Silva Study of structural, electrical, optical and magnetic properties of ZnO based films produced by magnetron sputtering" PhD thesis, university of Puerto Rico UPR

Auret, F., Meyer, W., Rensburg, P., Hayes, M., Nel, J., Wenckstern, H., Schmidt, H., Biehne, G., Hochmuth, H., Lorenz, M., Grundmann, M. (2007). Electronic properties of defects in pulsed-laser deposition grown ZnO with levels at 300 and 370meV below the conduction band. *Physica B*, Vol.401-402, pp. 378-381

Banerjee, A., Ghosh, C., Chattopadhyay, K., Minoura, H., Sarkar, A., Akiba, A., Kamiya, A., & Endo, T. (2006). Low-temperature deposition of ZnO thin films on PET and glass substrates by DC-sputtering technique. *Thin Solid Films*, Vol.496, pp. 112-116

Banerjee, A., Ghosh, C., Chattopadhyay, K., Minoura, H., Sarkar, A., Akiba, A., Kamiya, A., & Endo, T. (2006). Low-temperature deposition of ZnO thin films on PET and glass substrates by DC-sputtering technique. *Thin Solid Films*, Vol.496, pp. 112-116

Bang, K.-H., Hwang, D.-K., & Myoung, J.-M. (2003). Effects of ZnO buffer layer thickness on properties of ZnO thin films deposited by radio-frequency magnetron sputtering. *Appl. Surf. Sci.*, Vol.207, pp. 359-364

Basak, D., Amin, G., Mallik, B., Paul, G., & Sen, S. (2003). Photoconductive UV detectors on sol-gel-synthesized ZnO films. *J. Cryst. Growth*, Vol.256, pp. 73-77

Brabec, C., Sariciftci, N., & Hummelen, J. (2001). Plastic Solar Cells. *Adv. Funct. Mater.*, Vol.11, No.1, pp. 15-26

Brillson, L., Mosbacker, H., Hetzer, M., Strzhemechny, Y., Look, D., Cantwell, G., Zhang, J., & Song, J. (2008). Surface and near-surface passivation, chemical reaction, and Schottky barrier formation at ZnO surfaces and interfaces. *Appl. Surf. Sci.*, Vol.254, pp. 8000-8004

Brillson, L., Mosbacker, H., Hetzer, M., Strzhemechny, Y., Look, D., Cantwell, G., Zhang, J., & Song, J. (2008). Surface and near-surface passivation, chemical reaction, and Schottky barrier formation at ZnO surfaces and interfaces. *Appl. Surf. Sci.*, Vol.254, pp. 8000–8004

Caria, M., Barberini, L., Cadeddu, S., Giannattasio, A., Lai, A., Rusani, A., & Sesselego, A. (2001). Far UV responsivity of commercial silicon photodetectors. *Nucl. Instrum. Meth. A*, Vol.466, No.1, pp. 115-118

Chakraborty, A., Mondal, T., Bera, S., Sen, S., Ghosh, R., & Paul, G. (2008). Effects of Aluminium and Indium incorporation on the structural and optical properties of ZnO thin films synthesized by spray pyrolysis. *Mater. Chem. Phys.*, Vol.112, pp. 162-166

Chang, S., Chang, S., Chiou, Y., Lu, C., Lin, T., Lin, Y., Kuo, C., & Chang, H. (2007). ZnO photoconductive sensors epitaxially grown on sapphire substrates. *Sensor. Actuator., A*, Vol.140, pp. 60-64

Cho, H., Leerungnawarat, P., Hays, D., Pearton, S., Chu, S., Strong, R., Zetterling, C.-M., Östling, M., & Ren, F. (2000). low-damage dry etching of SiC. *Appl. Phys. Lett.*, Vol.76, No.6, pp. 739-741

Coppa, B., Fulton, C., Kiesel, S., Davis, R., Pandarinath, C., Burnette, J., Nemanich, R., & Smith, D. (2005). Structural, microstructural, and electrical properties of gold films and Schottky contacts on remote plasma-cleaned, n-type ZnO{0001} surfaces. *J. Appl. Phys.*, Vol.97. pp. 103517-1-13

Coppa, B., Fulton, C., Kiesel, S., Davis, R., Pandarinath, C., Burnette, J., Nemanich, R., & Smith, D. J. (2005). Structural, microstructural, and electrical properties of gold films and Schottky contacts on remote plasma-cleaned, n-type ZnO{0001} surfaces. *J. Appl. Phys.* Vol.97, pp. 103517-1-13

Craciun, V., Elders J., Gardeniers, J., & Boyd, I. (1994). Characteristics of high quality ZnO thin films deposited by pulsed laser deposition. *Appl. Phys. Lett.*, Vol. 65, No.23, pp. 2963-2965

Daraee, M., Hajian, M., Rastgoo, M., & Lavasanpour, L. (2008). Study of electrical characteristic of surface barrier detector with high series resistance. *Adv. Studies Theor. Phys.*, Vol.2, No.20, pp. 957-964.

Dong, Y. & Brillson, L. (2008). First-principles studies of metal (111)/ZnO {0001} interfaces. *J. Eectron. Mater.*, Vol.37, No.5, pp. 743-748

Ergin, B., Ketenci, E., & Atay, F. (2009). Characterization of ZnO films obtained by ultrasonic spray pyrolysis technique. *Int. J. Hydrogen Energ.*, Vol.34, pp. 5249-5254

Gao, W., & Li, Z. (2004). ZnO thin films produced by magnetron sputtering. *Ceram. Int.*, Vol.30, pp. 1155-1159

Ghosh, R., Basak, D., & Fujihara, S. (2004). Effect of substrate-induced strain on the structural, electrical, and optical properties of polycrystalline ZnO thin films. *J. Appl. Phys.*, Vol.96, No.5, pp. 2689-2692

Gökkavas, M., Butun, S., Tut, T., Biyikli, N., & Ozbay, E. (2007). AlGaN-based high-performance metal–semiconductor–metal photodetectors. *PNFA.*, Vol.5, pp. 53-62

Gümüş, C., Ozkendir, O., Kavak, H., & Ufuktepe, Y. (2006). Structural and optical properties of zinc oxide thin films prepared by spray pyrolysis method. *J. Optoelectron. Adv. M.*, Vol.8, No.1, pp. 299-303

Gür, E., Tüzemen, S, Kılıç, B., & Coşkun, C. (2007). High-temperature Schottky diode characteristics of bulk ZnO. *J. Phys.: Condens. Matter.*, Vol.19, pp. 196206-196214

Haas, F. (1997). Principles of Semiconductor Devices. RL-TR-96-217, In-House Report

Hanzaz, M., Bouhdada, A., Vigué, F., & Faurie, J. (2007). ZnSe-and GaN-based Schottky barrier photodetectors for blue and ultraviolet detection. *J. Act. Pass. Electron. Dev.*, Vol.2, pp. 165-169

Hickernell, F. (1976). Zinc-Oxide Thin-film surface-wave transducers. *Proceedings of the IEEE*, Vol. 64, No.5, pp. 631-635

Hiramatsu, K., & Motogaito, A. (2003). GaN-based Schottky barrier photodetectors from near ultraviolet to vacuum ultraviolet (360-50 nm). *phys. stat. sol. (a)*, Vol.195, No.3, pp. 496-501

Hussain, S. (2008). "Investigation of Structural and Optional Properties of Nanpcrystalline ZnO" PhD thesis, Department of Physics, Chemistry and Biology Linköping University

Hwang, K-S. , Kang, B-A., Jeong, J-H. , Jeon Y-S. , & Kim B-H. (2007). Spin coating-pyrolysis derived highly c-axis-oriented ZnO layers pre-fired at various temperatures. *Curr. Appl. Phys.*, Vol.7, pp. 421-425

Ianno, N., McConville, L., Shaikh, N., Pittal, S., & Snyder, P. (1992). Characterization of pulsed laser deposited Zinc oxide. *Thin Solid Films*, Vol.220, pp. 92- 99

Ip, K., Khanna, R., Norton, D., Pearton, S., Ren, F., Kravchenko, I., Kao, C., & Chi, G. (2005). Thermal stability of W2B and W2B5 contacts on ZnO. *Appl. Surf. Sci.*, Vol.252, pp. 1846-1853

Jandow N. N., F. K. Yam, S. M. Thahab, H. Abu Hassan, & Ibrahim, K. (2010a). Characteristics of ZnO MSM UV photodetector with Ni contact electrodes on Poly Propylene Carbonate (PPC) plastic substrate. *Curr. Appl. Phys.*, Vol.10, pp. 1452-1455

Jandow, N., Ibrahim, K., Abu Hassan, H. (2010b). I-V Characteristic for ZnO MSM Photodetector with Pd Contact Electrodes on PPC Plastic, *AIP Conf. Proc.*, Vol.1250, pp. 424-427

Jandow, N., Ibrahim, K., Abu Hassan, H., Thahab, S., & Hamad, O. (2010c). The electrical properties of ZnO MSM Photodetector with Pt Contact Electrodes on PPC Plastic. *JED*, Vol.7, pp. 225-229

Jandow, N., Ibrahima, K., Yam, F., Abu Hassan, H., Thahab, S., & Hamad, O. (2010d). The study of ZnO MSM UV photodetector with Pd contact electrodes on (PPC) plastic. *J.Optoelectron. Adv. Mat.-Rabid*, Vol.4, No.5, pp. 726-730

Jandow, N., Yam, F., Thahab, S., Ibrahim, K., Abu Hassan, H. (2010). The characteristics of ZnO deposited on PPC plastic substrate. *Mater. Lett.*, Vol.64, pp. 2366-2368

Janotti, A., & Van de Walle, C. (2009e). Fundamentals of zinc oxide as a semiconductor. *Rep. Prog. Phys.*, Vol. 72, pp. 126501-126530

Jeong, I.-S., Kim, J., Park, H.-Ho, & Im, S. (2004). n-ZnO/p-Si UV photodetectors employing AlO$_x$ films for antireflection. *Thin Solid Films*, Vol.447-448, pp. 111-114

Jiang, D., Zhang, J., Lu, Y., Liu, K., Zhao, D., Zhang, Z., Shen, D., & Fan, X. (2008) Ultraviolet Schottky detector based on epitaxial ZnO thin film. *Solid State Electron.*, Vol.52, pp. 679-682

Jin, B., Woo, H., Im, S., Bae, S., & Lee, S. (2001). Relationship between photoluminescence and electrical properties of ZnO thin films grown by pulsed laser deposition. *Appl. Surf. Sci.*, Vol.169-170, pp. 521-524

Jin, C. (2003). "Growth and characterization of ZnO and ZnO-based Alloys-MgxZn1-xO and MnxZn1-xO". PhD thesis, Department of materials science and Engineering, North Carolina state university, Raleigh

Jun, W., Degang, Z., Zongshun, L., Gan F., Jianjun, Z., Xiaomin, S., Baoshun, Z., & Hui, Y. (2003). Metal-semiconductor-metal ultraviolet photodetector based on GaN. *Sci. China Ser. G*, Vol.46, No.2, pp. 198-203

Kahng, D. (1963). Conduction properties of the Au-n-type-Si Schottky barrier. *Solid State Electron.*, Vol. 6, pp. 281-295

Kang, S., Joung, Y., Chang, D., & Kim, K. (2007). Piezoelectric and optical properties of ZnO thin films deposited using various $O_2/(Ar+O_2)$ gas ratios. *J. Mater. Sci: Mater Electron.*, Vol.18, pp. 647-653

Khoury, A., al Asmar, R., Abdallah, M., El Hajj Moussa, G., & Foucaran, A. (2010). Comparative study between zinc oxide elaborated by spray pyrolysis, electron beam evaporation and rf magnetron techniques. *Phys. Status Solidi A*, Vol.207, No.8, pp. 1900-1904

Kim, H., Kim, H., & Kim, D.-W. (2010). Silver Schottky contacts to a-plane bulk ZnO. *J. Appl. Phy.*, Vol.108, pp. 074514-1-5

Kim, J., Lee J., Lee, J., Lee, D., Jang, B., Kim, H., Lee W., Cho, C., & Kim, J. (2009). Characteristics of ZnO films deposited on plastic substrates at various rf sputtering powers. *J. Korean Phys. Soc.*, Vol.55, No.5, pp. 1910-1914

Kim, J., Yun, J., Kim, C., Park, Y., Woo, J., Park, J., Lee, J., Yi, J., & Han, C. (2010). ZnO nanowire-embedded Schottky diode for effective UV detection by the barrier reduction effect. *Nanotechnology*, Vol.21, pp. 115205-115210

Kiriakidis, G., Suchea, M., Christoulakis, S., Horvath, P., Kitsopoulos, T., & Stoemenos, J. (2007). Structural characterization of ZnO thin films deposited by dc magnetron sputtering. *Thin Solid Films*, Vol.515, pp. 8577-8581

Koch, U., Fojtik, A., Weller, H., & Henglein, A. (1985). Photochemistry of semiconductor colloids, preparation of extremely small ZnO particles, fluorescence phenomena and size quantization effects, *Chem.Phys.Lett.*, Vol.122, pp. 507-510

Kumar, S., Kim, H.., Sreenivas, K., & Tandon, P. (2009). ZnO based surface acoustic wave ultraviolet photo sensor. *J. Electroceram*, Vol.22, pp. 198-202

Lai, L.-W., & Lee, C.-T. (2008). Investigation of optical and electrical properties of ZnO thin films, *Mater. Chem. Phys.*, Vol.110, pp. 393-396

Lakhotia, G., Umarji, G., Jagtap, S., Rane, S., Mulik, U., Amalnerkar, D., & Gosavi, S. (2010). An investigation on TiO_2-ZnO based thick film 'solar blind', photo-conductor for 'green' electronics. *Mater. Sci. Eng. B*, Vol.168, pp. 66-70

Liang, S., Sheng, H., Liu, Y., Huo, Z., Lu, Y., & Shen, H. (2001). ZnO Schottky ultraviolet photodetectors. *J. Cryst. Growth*, Vol.225, pp. 110-113

Lim, J., Kang/ Ch., Kim, K., Park, I., Hwang, D., & Park, S. (2006). UV Electroluminescence Emission from ZnO light-emitting diodes grown by high-temperature radiofrequency sputtering. *Adv. Mater.*, Vol.18, pp. 2720-2724

Lin, T., Chang, S., Su, Y., Huang, B., Fujita, M., & Horikoshi, Y. (2005). ZnO MSM photodetectors with Ru contact electrodes. *J. Cryst. Growth*, Vol.281, pp. 513-517

Liu, K., Ma, J., Zhang, J., Lu, Y., Jiang, D., Li, B., Zhao, D., Zhang, Z., Yao, B., & Shen, D. (2007). Ultraviolet photoconductive detector with high visible rejection and fast photoresponse based on ZnO thin film. *Solid State Electron.*, Vol.51, pp. 757-761

Liu, K., Sakurai, M., & Aono, M. (2010). ZnO-based ultraviolet photodetectors. *Sensors*, Vol.10, pp. 8605-8630

Liu, Y., Egawa, T., Jiang, H., Zhang, B., Ishikawa, H., & Hao, M. (2004). Near-ideal Schottky contact on quaternary AlInGaN epilayer lattice-matched with GaN, *Appl. Phys. Lett.*, Vol.85, No.24, pp. 6030-6032

Liu, Y., Gopla, C., Liang, S., Emanetoglu, N., Lu, Y., Shen, H., & Wraback, M. (2000). Ultraviolet detectors based on epitaxial ZnO films grown by MOCVD. *J. Electron. Mater*, Vol.29, No.1, pp. 69-74

Liu, Y., Li, Q., & Shao, H. (2009). Properties of ZnO-Al films deposited on polycarbonate substrate. *Vacuum*, Vol.83, pp. 1435–1437

Liu, Y.-Y., Yuan, Y. –Z., Gao, X.-T., Yan, S.-S., Cao, X.-Z., & Wei, G.-X. (2007). Deposition of ZnO thin film on polytetrafluoroethylene substrate by the magnetron sputtering method. *Mater. Lett.*, Vol.61, pp. 4463-4465

Lu, X., Zhu, Q., & Meng, Y. (2005). Kinetic analysis of thermal decomposition of Poly(Propylene Carbonate), *Polym. Degrad. Stab.*, Vol.89, pp. 282-288

Lu, Y.-M., Tsai, S.-Y., Lu, J.-J., & Hon, M.-H. (2007). The structural and optical properties of zinc oxide thin films deposited on PET substrate by rf. magnetron sputtering. *Solid State Phenom.*, Vol.121-123, pp. 971-974

Ma, C., Taya, M., & Xu, C. (2008). Flexible electrochromic device based on poly (3, 4-(2, 2-dimethylpropylenedioxy) thiophene). *Electrochim. Acta*, Vol.54, pp. 598-605

Mahmood, F., Gould, R., Hassan, A., & Salih, H. (1995). DC. properties of ZnO thin films prepared by rf. magnetron sputtering. *Thin Solid Films*, Vol.770, pp. 376-379

Mandalapu, L., Xiu, F., Yang, Z., & Liu, J. (2007). Ultraviolet photoconductive detectors based on Ga-doped ZnO films grown by molecular-beam epitaxy. *Solid State Electron.*, Vol.51, pp. 1014-1017

Mead, C. (1966). Metal-semiconductor surface barriers. *Solid State Electron.*, Vol.9, pp. 1023-1033

Mead, C., & Spitzer, W. (1963). Fermi level position at semiconductor surfaces. *Phys. Rev. Lett.*, Vol.10, No.11, pp. 471-472

Monroy, E., Omnés, F., & Calle, F. (2003). Wide-bandgap semiconductor ultraviolet photodetectors. *Semicond. Sci. Technol.*, Vol.18, pp. R33-R51

Moon, T.-H., Jeong, M.-C., Lee, W., & Myoung, J.-M. (2005). The fabrication and characterization of ZnO UV detector. *Appl. Surf. Sci.*, Vol.240, pp. 280-285

Myoung, J.-M., Yoon, W.-H., Lee, D.-H., Yun, I., Bae, S.-H., & Lee, S.-Y. (2002). Effects of thickness variation on properties of ZnO thin films grown by pulsed laser deposition, *Jpn. J. Appl. Phys.*, Vol.41, pp. 28-31

Nandy, S., Goswami, S., & Chattopadhyay, K. (2010). Ultra smooth NiO thin films on flexible plastic (PET) substrate at room temperature by rf magnetron sputtering and effect of oxygen partial pressure on their properties, *Appl. Surf. Sci.*, Vol.256, pp. 3142-3147

Nause, J., & Nemeth, B. (2005). Pressurized melt growth of ZnO boules. *Semicond. Sci. Technol.*, Vol.20, pp. S45-S48

Nav Bharat Metallic Oxide Industries Pvt. Limited. n.d. *Some of the popular applications of Zinc Oxide.* Available from: <http://www.navbharat.co.in/Clients.htm>

Neville, R., & Mead, C. (1970). "Surface barriers on zinc oxide". *J. Appl. Phys.*, Vol.41, No.9, pp. 3795- 3800

Nunes, P., Fortunato, E., & Martins, R. (2001). Influence of the annealing conditions on the properties of ZnO thin films. *Int. J. Inorg. Mater.*, Vol.3, pp. 1125-1128

Ott, A., & Chang, R. (1999). Atomic layer-controlled growth of transparent conducting ZnO on plastic substrates. *Mater. Chem. Phys.*, Vol.58, pp. 132-138

Paul, G., Bhaumik, A., Patra, A., & Bera, S. (2007). Enhanced photo-electric response of ZnO/polyaniline layer-by-layer self-assembled films, *Mater. Chem. Phys.*, Vol.106, pp. 360-363

Plaza, J., Martínez, O., Carcelén, V., Olvera, J., Sanz, L., & Diéguez, E. (2008). Formation of ZnO and Zn1-xCdxO films on CdTe/CdZnTe single crystals. *Appl. Surf. Sci.*, Vol.254, pp. 5403-5407

Polyakov, A., Smirnov, N., Kozhukhova, E., Vdovin, V., Ip, K., Heo, Y., Norton, D., & Pearton, S. (2003). "Electrical characteristics of Au and Ag Schottky contacts on n-ZnO". *Appl. Phys. Lett.*, Vol.83, No.8, pp. 1575-1577

Rabadanov, R., Guseikhanov, M., Aliev, I., & Semiletov, S. (1981). Properties of metal-zinc oxide contacts. *Russ. Phys. J.*, Vol.24, No.6, pp. 548-551

Rabadanov, R., Guseikhanov, M., Aliev, I., & Semiletov, S. (1982). Properties of metal-zinc oxide contacts. *Russ Phys J.*, Vol.24, No.6, pp. 548-551

Rao, T., Kumar, M., Safarulla, A., Ganesan, V., Barman, S., & Sanjeeviraja, C. (2010). Physical properties of ZnO thin films deposited at various substrate temperatures using spray pyrolysis. *Physica B*, Vol.405, pp. 2226-2231

Roccaforte, F., Giannazzo, F., Iucolano, F., Eriksson, J., Weng, M., & Raineri, V. (2010). Surface and interface issues in wide band gap semiconductor electronics. *Appl. Surf. Sci.*, Vol.256, pp. 5727-5735

Sans, J., Segura, A., Mollar, M., & Marí, B. (2004). Optical properties of thin films of ZnO prepared by pulsed laser deposition. *Thin Solid Films*, Vol.453-454, pp. 251-255

Shan, F., Liu, G., Lee, W., Lee, G., Kim, I., Shin, B., & Kim, Y. (2005). Transparent conductive ZnO thin films on glass substrates deposited by pulsed laser deposition. *J. Cryst. Growth*, Vol.277, pp. 284-292

Shan, F., Shin, B., Jang, S., & Yu, Y. (2004). Substrate effects of ZnO thin films prepared by PLD technique. *J. Eur. Ceram. Soc.*, Vol.24, pp. 1015-1018

Sofiani, Z., Derkowska, B., Dalasiński, P., Wojdyła, M., Dabos-Seignon, S., Alaoui Lamrani, M., Dghoughi, L., Bała, W., Addou, M., Sahraoui, B. (2006). Optical properties of ZnO and ZnO:Ce layers grown by spray pyrolysis. *Opt. Commun.*, Vol.267, pp. 433-439

Song, P.-f., Wang, S.-j., Xiao, M., Du, F.-g., Gan, L.-q., Liu, G.-q., & Meng, Y.-z. (2009). Cross-linkable and thermally stable aliphatic polycarbonates derived from CO_2, propylene oxide and maleic anhydride. *J Polym Res*, Vol.16, pp. 91-97

Sze, S. (2002). Semiconductor devices, physics and technology, 2nd ed., John Wiley & Sons New York , USA

Sze, S., & Kwok K. (2006). Physics of semiconductor devices, 3rd ed., United States of America

Sze, S., Coleman, D., & Loya, J., (1971). Current transport in metal-semiconductor-metal (MSM) structures. *Solid State Electron.*, Vol.14, pp. 1209-1218

Tan, S., Chen, B., Sun, X., Fan, W., Kwok, H., Zhang, X., & Chua, S. (2005). Blueshift of optical band gap in ZnO thin films grown by metal-organic chemical-vapor deposition. *J. Appl. Phys.*, Vol.98, pp. 013505-013505-5

Tneh, S., Hassan, Z., Saw, K., Yam, F., Abu Hassan, H. (2010). The structural and optical characterizations of ZnO synthesized using the "bottom-up" growth method. *Physica B*, Vol.405, pp. 2045-2048

Tsai, H-Y. (2007). Characteristics of ZnO thin film deposited by ion beam sputter. *J. Materi. Process. Technol.*, Vol.192-193, pp. 55-59

Tsai, S.-Y., Lu, Y.-M., Lu, J.-J., & Hon, M.-H. (2006). Comparison with electrical and optical properties of zinc oxide films deposited on the glass and PET substrates. *Surf. Coat. Technol.*, Vol.200, pp. 3241-3244

Tung, R. (2001). Recent advances in Schottky barrier concepts. *Mater. Sci. Eng., R*, Vol.35, pp. 1-138

Tüzemen, S., & Gür, E. (2007). Principal issues in producing new ultraviolet light emitters based on transparent semiconductor zinc oxide. *Opt. Mater.*, Vol.30, pp. 292-310

Vanlaar, J. & Scheer, J. (1965). Fermi level stabilization at semiconductor surfaces. *Surf. Sci.* Vol.3, pp. 189-201

Wang, Q., Pflügl, C., Andress, W., Ham, D., & Capasso, F. (2008). Gigahertz surface acoustic wave generation on ZnO thin films deposited by radio frequency magnetron sputtering on III-V semiconductor substrates. *J. Vac. Sci. Technol. B*, Vol.26, No.6, pp. 1848-1851

Wei, X., Zhang, Z., Liu, M., Chen, C., Sun, G., Xue, C., Zhuang, H., & Man, B. (2007) Annealing effect on the microstructure and photoluminescence of ZnO thin films. *Mater. Chem. Phys.*, Vol.101, pp. 285-290

Wright, J., Stafford, L., Gila, B., Norton, D., Pearton, S., Wang, H.-T., & Ren, F. (2007). Effect of cryogenic temperature deposition of various metal contacts on bulk single-crystal n-type ZnO, *J. Electron. Mater.*, Vol.36, No.4, pp. 488-493

Wu, L., Tok, A., Boey, F., Zeng, X., & Zhang, X. (2007). Chemical Synthesis of ZnO Nanocrystals. *IEEE. Trans. Nanotechnol*, Vol.6, No.5, pp. 497-503

Yam, F., & Hassan, Z. (2008). The investigation of dark current reduction in MSM photodetector based on porous GaN. *J. Optoelectron Adv. M.*, Vol.10, pp. 545-548

Yan, F., Xin, X., Aslam, S., Zhao, Y., Franz, D., Zhao, J., & Weiner, M. (2004). 4H-SiC UV Photo detectors with large area and very high specific detectivity. *IEEE J. Quantum Elect.*, Vol.40, No.9, pp. 1315-1320

Yıldırım, N., Ejderha, K., & Turut, A. (2010). On temperature-dependent experimental I-V and C-V data of Ni/n-GaN Schottky contacts. *J. Appl. Phys.*, Vol.108, pp. 114506-1-8

Young, S., Ji, L., Chang, S., & Su, Y. (2006). ZnO metal–semiconductor–metal ultraviolet sensors with various contact electrodes. *J. Cryst. Growth*, Vol.293, pp. 43-47.

Young, S., Ji, L., Chang, S., Chen, Y., Lam, K., Liang, S., Du, X., Xue, Q., & Sun, Y. (2007). ZnO metal-semiconductor-metal ultraviolet photodetectors with Iridium contact electrodes. *IET Optoelectron.*, Vol.1, No.3, pp. 135-139

Young, S., Ji, L., Fang, T., Chang, S., Su, Y., & Du X. (2007). ZnO ultraviolet photodiodes with Pd contact electrodes. *Acta Mater.*, Vol.55, pp. 329-333

Zhang, B., Wakatsuki, K., Binh, N., Usami, N., & Segawa, Y. (2004). Effects of growth temperature on the characteristics of ZnO epitaxial films deposited by metalorganic chemical vapor deposition. *Thin Solid Films*, Vol.449, pp. 12-19

Zhang, J., Kang, J., Hu, P., & Meng, Q. (2007). Surface modification of poly(propylene carbonate) by oxygen ion implantation. *Appl. Surf. Sci.*, Vol.253, pp. 5436-5441

2

Silicon Photodetectors Based on Internal Photoemission Effect: The Challenge of Detecting Near-Infrared Light

Maurizio Casalino, Luigi Sirleto, Mario Iodice and Giuseppe Coppola

Istituto per la Microelettronica e Microsistemi, Consiglio Nazionale delle Ricerche, Naples

Italy

1. Introduction

Silicon Photonics has emerged as an interesting field due to its potential for low-cost optical components integrated with electronic functionality. In the past two decades, there has been growing interest in photonic devices based on Si-compatible materials (Kimerling et al., 2004; Jalali & Fathpour, 2006) in the field both of the optical telecommunications and of the optical interconnects. In this contest, tremendous progresses in the technological processes based on the use silicon-on insulator (SOI) substrates have allowed to obtain reliable and effectiveness full complementary metal-oxide semiconductor (CMOS) compatible optical components such as, low loss waveguides, high-Q resonators, high speed modulators, couplers, and optically pumped lasers (Rowe et al., 2007 ; Vivien et al., 2006 ; Xu et al., 2007 ; Michael et al., 2007; Liu et al., 2007 ; Liu et al., 2006). All these devices have been developed to operate in the wavelength range from C optical band (1528 – 1561 nm) to L optical band (1561 – 1620 nm). However one of the crucial steps toward the integration of photonics with electronics resides in the development of efficient chip-scale photodetectors (PD) integrated on Si. Bulk photodetectors are perhaps the oldest and best understood silicon optoelectronic devices. Commercial products in Si operate at wavelengths below 1100 nm, where band-to-band absorption occurs. For the realization of photodiodes integrated in photonics circuits operating at wavelengths beyond 1100 nm silicon is not the right material because its transparency. In the last years, in order to take advantage of low-cost standard Si-CMOS processing technology, a number of photodetectors have been proposed based on different physical effects, such as: defect-state absorption (Bradley et al., 2005), two photon absorption (TPA) (Liang et al., 2002) and internal photoemission absorption (Zhu et al., 2008a). Physical effects, working principles, main structures reported in literature and the most significant results obtained in recent years were reviewed and discussed in our previous paper (Casalino et al., 2010a). In this paragraph, we go into more depth on photodetectors based on the internal photoemission effect (IPE). Silicon infrared photodiodes based on IPE are not novel, in fact PtSi/p-Si, Ir/p-Si and Pd$_2$Si/p-Si junctions are usually used in the infrared imaging systems (Kosonocky et al., 1985). The main advantages of these devices resides in their extremely high switching speed and in their simple fabrication process, but, due to high background current density these devices can only work at cryogenic temperature.

However, in recent years IPE has emerged as a new option for detecting also near infrared (NIR) wavelengths at room temperature. Unfortunately, the photoemission quantum efficiency is low compared to that of detectors based on inter-band absorption and this limits the application both in power monitoring and in the telecommunication field. Low quantum efficiency is a direct result of conservation of momentum during carrier emission over the potential barrier, in fact, the majority of excited carriers which do not have enough momentum normal to the barrier are reflected and not emitted. Moreover as incoming photons can excite carriers lying in states far below the Fermi energy, which can not overcome over the metal-semiconductor potential barrier, the quantum efficiency of these devices is further decreased. In recent years new approaches and structures are proposed in order to circumvent these limitations.

In this paragraph an overview of the state of the art of NIR all-Si photodetectors based on IPE is presented. First, the physical effects of IPE and the main figures of merit of devices based on IPE are elucidated. Then, the main structures reported in literature, starting from historical devices for imaging application at infrared wavelengths, up to new NIR devices which could be adapted for telecommunications and power monitoring, are described in detail. Finally the most significant results obtained in the last years are reviewed and discussed comparing the performances of devices based on different approaches.

2. IPE-based device performances

We hereafter present a theoretical background useful to clarify the physics behind the working principle of devices and we analyze crucial points affecting device performances.

2.1 Device efficiency and IPE theory

IPE is the optical excitation of electrons in the metal to energy above the Schottky barrier and then transport of these electrons to the conduction band of the semiconductor. The standard IPE theory is due to Fowler (Fowler, 1931). However, the Fowler's theory was obtained without taking into account the thickness of the Schottky metal layer. In 1973, Archer and Cohen (Archer & Cohen, 1973) proposed thinning the electrode to increase the emission probability. The enhancement of IPE in thin metal film was theoretically investigated by Vicker who introduce a moltiplicative factor to the Fowler's formula (Vickers, 1971). However the resulting electrode was so thin that it was semitransparent producing low metal absorbance. More recently Casalino et al. proposed to improve the absorbance in NIR photodetector based on IPE by using a microcavity Fabry-Perot (Casalino et al., 2008a, 2010b). Finally a further enhancement of IPE can be obtained due to image force effect which modify the Schottky barrier under a reverse bias favouring IPE. This effect can be taken into account by adding a barrier collection efficiency term. The resulting quantum efficiency of a photodetector based on IPE, can be written by the formula (Casalino et al., 2008b):

$$\eta = A_T F_e P_E \eta_c \qquad (1)$$

where A_T is the total optical absorbance of the metal, F_e is the Fowler' factor, P_E is the Vicker's factor and η_c is the bias dependent barrier collection efficiency. Very often, device efficiency is described in term of responsivity, a very important property of a detector indicating the current produced (I_{ph}) by a certain optical power (P_{OPT}). Responsivity is strictly linked to a device's quantum efficiency by the formula:

$$\Re = \frac{I_{ph}}{P_{OPT}} = \frac{\lambda[nm]}{1242}\eta \qquad (2)$$

Reasonable responsivities are necessary for an acceptable signal-to-noise ratio and to ease the design and realization of the amplifier circuitry that follows.

2.1.1 Standard fowler's theory of IPE

IPE is the optical excitation of electrons into the metal to an energy above the Schottky barrier and then transport of these electrons to the conduction band of the semiconductor (Fig. 1).

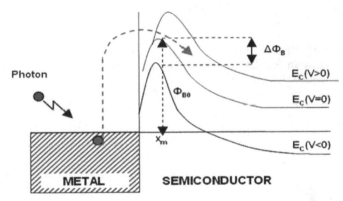

Fig. 1. Energy band diagram for a metal/n-semiconductor junction. "Reprinted with permission from M. Casalino *et al.*, "A silicon compatible resonant cavity enhanced photodetector working at 1.55 μm," Semicond. Sci. Technol., 23, 075001, 2008 (doi: 10.1088/0268-1242/23/7/075001). IOP Publishing is acknowledged."

The standard theory of photoemission from a metal into the vacuum is due to Fowler (Fowler, 1931). In a gas of electrons obeying the Fermi-Dirac statistic, if photon energy is close to potential barrier ($h\nu \approx \Phi_B$), the fraction (F_e) of the absorbed photons, which produce photoelectrons with the appropriate energy and momenta before scattering to contribute to the photocurrent, is given by:

$$F_e = \frac{\left[(h\nu - (\phi_{B0} - \Delta\phi_B))^2 + \frac{(kT\pi)^2}{3} - 2(kT)^2 e^{-\frac{h\nu - (\phi_{B0} - \Delta\phi_B)}{kT}} \right]}{8kTE_F \log\left[1 + e^{-\frac{h\nu - (\phi_{B0} - \Delta\phi_B)}{kT}} \right]} \qquad (3)$$

where hv is photons energy, Φ_{B0} is the potential barrier at zero bias, $\Delta\Phi_B$ is the lowering due to image force effect (as we will see later), E_F is the metal Fermi level, k is the Boltzmann's constant and T is the absolute temperature. As it is possible to see in Eq. 3, F_e is strongly depending from the potential barrier height of the metal-semiconductor interface.

2.1.2 IPE enhancement in thin metal films

In order to study the quantum efficiency for thin metal films, the theory must be further extended, taking into account multiple reflections of the excited electrons from the surfaces of the metals film, in addition to collisions with phonons, imperfections and cold electrons. Assuming a thin metal film, a phenomenological, semiclassical, ballistic transport model for the effects of the scattering mechanisms resulting in a multiplicative factor for quantum efficiency was developed by Vickers (Vickers, 1971) and recently reviewed by Scales *et al.* (Scales & Berini, 2010). According to this model the accumulated probability P_E that the electrons will have sufficient normal kinetic energy to overcome potential barrier is given by:

$$P_E \cong \frac{L_e}{d}\left[1 - e^{-\frac{d}{L_e}}\right]^{\frac{1}{2}} \tag{4}$$

where d is the metal thickness and L_e the metal mean free path. For example, as it can be seen by plotting Eq. 4 in Fig. 2 (assuming a copper mean free path of 45 nm (Chan et al., 1980a), the lower metal thickness, the higher P_E.

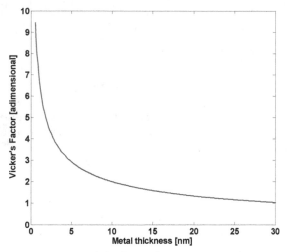

Fig. 2. Vicker's factor P_E versus metal thickness for copper metal.

It is worth noting that in a recent work (Scales & Berini, 2010), Scales and Berini show that a further enhancement of this probability emission can be obtained in structures realized with thin metal film buried in a semiconductor and forming two Schottky barriers.

2.1.3 IPE enhancement by optical cavity

The enhancement of IPE in thin metal film lead to a resulting electrode so thin to be semitransparent, producing low metal absorbance. In order to increase the absorbance in thin metal film, the use of a microcavity Fabry-Perot has been proposed. The sketch of proposed photodetector is shown in Fig. 3. The resonant cavity is a Fabry–Perot vertical to-the-surface structure. It is formed by a distributed Bragg reflector (DBR), a metallic top mirror and in between a Si cavity. A dielectric layer on top of the metal is generally considered for thin metal protection purpose.

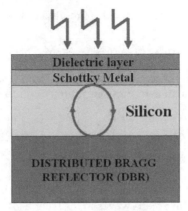

Fig. 3. Sketch of IPE photodetector based on optical cavity.

The proposed structure of Fig. 4 can be modelled by the multilayer shown in Fig. 3 and absorbance calculation can be carried out by the Transfer Matrix Method (TMM) (Muriel & Carballar, 1997).

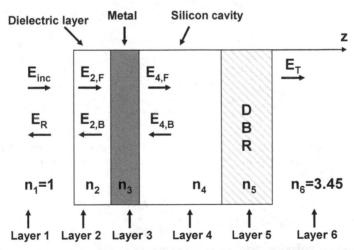

Fig. 4. Multilayer schematizing device in fig. 3 for absorbance calculation by TMM.

Normal incidence condition and the unidimensional variations of refractive index (n) along the propagation direction z, are taken into account. Let E_{inc} and E_R be the electric field complex amplitude of the incident and reflected waves at the interface between the air and the dielectric layer, and E_T the waves transmitted by final plane of structure, by TMM we obtain:

$$\begin{bmatrix} E_{inc} \\ E_R \end{bmatrix} = M_{TOT} \begin{bmatrix} E_T \\ 0 \end{bmatrix}$$

$$E_T = \frac{1}{M_{TOT_{11}}} E_{inc} \qquad E_R = \frac{M_{TOT_{21}}}{M_{TOT_{11}}} E_{inc}$$

(5)

where $M_{TOT_{ij}}$ are the element of the matrix $\mathrm{M_{TOT}}$ representing the whole multilayer:

$$M_{TOT} = M_{1/2} \cdot M_2 \cdot M_{2/3} \cdot M_3 \cdot M_{3/4} \cdot M_4 \cdot M_{4/5} \cdot M_5 \cdot M_{5/6}$$

(6)

where $\mathrm{M}_{i/j}$ and M_i are the interface matrix between i and j layer and the i-layer matrix, respectively, which can calculated by knowing refractive index (n_i) and thickness (d_i) of every i^{th} layer:

$$M_{i/j} = \frac{1}{2n_i} \begin{pmatrix} n_i + n_j & n_i - n_j \\ n_i - n_j & n_i + n_j \end{pmatrix} \qquad M_i = \begin{pmatrix} e^{jk_0 n_i d_i} & 0 \\ 0 & e^{-jk_0 n_i d_i} \end{pmatrix}$$

(7)

Let $E_{2,F}$ $(E_{4,F})$ and $E_{2,B}$ $(E_{4,B})$ be the frequency domain electric field complex amplitudes of the forward and backward travelling plane waves in the layer 2 (4), by TMM we obtain:

$$\begin{bmatrix} E_{2,F} \\ E_{2,B} \end{bmatrix} = M_A \begin{bmatrix} E_T \\ 0 \end{bmatrix}$$

$$E_{2,F} = \frac{M_{A_{11}}}{M_{TOT_{11}}} E_{inc} \qquad E_{2,B} = \frac{M_{A_{21}}}{M_{TOT_{11}}} E_{inc}$$

(8)

and

$$\begin{bmatrix} E_{4,F} \\ E_{4,B} \end{bmatrix} = M_B \begin{bmatrix} E_T \\ 0 \end{bmatrix}$$

$$E_{4,F} = \frac{M_{B_{11}}}{M_{TOT_{11}}} E_{inc} \qquad E_{4,B} = \frac{M_{B_{21}}}{M_{TOT_{11}}} E_{inc}$$

(9)

where $\mathrm{M_A}$ is the matrix calculated from interface between protection coating and metal to the final plane, while $\mathrm{M_B}$ is the matrix calculated from interface between metal and silicon cavity to the final plane,

$$M_A = M_{2/3} \cdot M_3 \cdot M_{3/4} \cdot M_4 \cdot M_{4/5} \cdot M_5 \cdot M_{5/6}$$
$$M_B = M_4 \cdot M_{4/5} \cdot M_5 \cdot M_{5/6}$$

(10)

The optical power is linked at the electrical field by:

$$P = \frac{n}{2\eta_0} |E|^2$$

(11)

where η_0 is the vacuum characteristic impedance of electromagnetic waves. The total power going in (P_{input}) and going out (P_{output}) from the metal is given by:

$$P_{input} = \frac{n_2}{2\eta_0}\left|E_{2,F}\right|^2 + \frac{n_4}{2\eta_0}\left|E_{4,B}\right|^2 = \left(n_2\left|\frac{M_{A_{11}}}{M_{TOT_{11}}}\right|^2 + n_4\left|\frac{M_{B_{21}}}{M_{TOT_{11}}}\right|^2\right)\frac{\left|E_{inc}\right|^2}{2\eta_0}$$

$$P_{output} = \frac{n_2}{2\eta_0}\left|E_{2,B}\right|^2 + \frac{n_4}{2\eta_0}\left|E_{4,F}\right|^2 = \left(n_2\left|\frac{M_{A_{21}}}{M_{TOT_{11}}}\right|^2 + n_4\left|\frac{M_{B_{11}}}{M_{TOT_{11}}}\right|^2\right)\frac{\left|E_{inc}\right|^2}{2\eta_0}$$

(12)

and metal absorbance can be written as (Casalino, 2006a, 2006b):

$$A_T = \frac{P_{input} - P_{output}}{P_{inc}} = \left\{\left(n_2\left|\frac{M_{A_{11}}}{M_{TOT_{11}}}\right|^2 + n_4\left|\frac{M_{B_{21}}}{M_{TOT_{11}}}\right|^2\right) - \left(n_2\left|\frac{M_{A_{21}}}{M_{TOT_{11}}}\right|^2 + n_4\left|\frac{M_{B_{11}}}{M_{TOT_{11}}}\right|^2\right)\right\}$$

(13)

where n_2 and n_4 are the refractive indices of the 2th and 4th layer, respectively.

In a work of 2006, Casalino *et al.* (Casalino et al., 2006a, 2006b) proposed a methodology in order to design an optimum device, in fact with a right choice of the multilayer thicknesses a maximum in absorbance can be obtained. In the device proposed by the authors DBR is considered formed by alternate layers of Si and SiO₂ having refractive index 3.45 and 1.45, and thickness of 340 nm and 270 nm, respectively, while a metallic layer of gold, having refractive index n_{Au}=0.174+j9.96 (Chan & Card, 1980b), mean free path L_e=55 nm (Chan et al., 1980a), Fermi level E_F=5.53 eV (Yeh, 1988) and thickness d=32 nm, was chosen as Schottky contact. We point out that in the simulations, semi-infinite first and last layers of air (n_1=1) and silicon (n_6=3.45), respectively, are considered. In Fig. 5 absorbance plotted against wavelengths is reported for the proposed device without and with DBR, showing a significant enhancement in absorbance, due to the optical cavity, of almost two order of magnitude at 1550 nm.

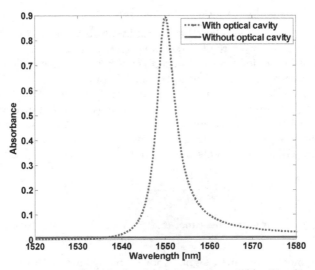

Fig. 5. Absorbance versus wavelengths for the device proposed by Casalino et al. (Casalino et al., 2006a, 2006b).

2.1.4 IPE enhancement by reverse voltage applied

In order to complete the transaction of the IPE theory, the image force between an electron and the metal surface must be taken into account. Due to the image force effect a lowering ($\Delta\Phi_B$) and displacement (x_m) of the metal-semiconductor interface potential barrier, is provided. These barrier lowering and displacement are given by (Sze, 1981):

$$\Delta\phi_B = \sqrt{\frac{q}{4\pi\varepsilon_{Si}}\frac{|V_{Bias}|}{W}} \qquad x_m = \sqrt{\frac{q}{16\pi\varepsilon_{Si}}\frac{W}{|V_{Bias}|}} \tag{14}$$

where ε_{Si} is the permittivity of silicon (10^{-12} C/cmV), W is the depletion width and V_{Bias} the applied bias voltage.

It is worth noting that while the potential barrier lowering can be taken into account by Fowler's theory (Eq. 3), the potential barrier displacement influences the probability that an electron undergoes scattering phenomena in Si travelling from the metal-semiconductor interface to the Schottky barrier maximum. This probability can be taken into account by the barrier collection efficiency (η_c):

$$\eta_c = e^{-\frac{x_m}{L_s}} \tag{15}$$

where L_s is the electron scattering length in the silicon. It is worth noting that increasing the bias voltage, a shift of Schottky barrier closer to metal/semiconductor interface is obtained. Therefore, the barrier collection efficiency increases. The improvement of the device efficiency by increasing reverse voltage has been theoretically investigated by Casalino et al. (Casalino et al., 2008b).

2.2 Bandwidth of IPE-based devices

The electrical properties of diodes based on Schottky junctions are determined by majority carrier phenomena, while for p-n diodes they are primarily determined by minority carriers. Therefore, the Schottky diodes can be switched faster because there are no minority carrier storage effects. The response time of Schottky barrier photodiodes can be determined by three parameters: the diffusion time in the quasi neutral region; the electrical frequency response or RC time required to discharge the junction capacitance through the resistance and the transit time across the depletion region. By designing the diode in such a way that depleted region length (W) equals to the device length (L), under suitable revere bias applied, the diffusion time can be neglected and the total frequency (f_{tot}) can be written as:

$$\frac{1}{f_{tot}} = \frac{1}{f_{tr}} + \frac{1}{f_{RC}} \tag{16}$$

where f_{tr} and f_{RC} are the transit time and time constant limited 3-dB bandwidth, respectively.

According to the model for small signal shown in Fig. 6 (Casalino et al., 2010b), the photodetector can be schematized as current generator (I_{ph}), resistance (R_j), capacitance (C_j) associated to the junction, series resistance (R_s) and load resistance (R_{load})

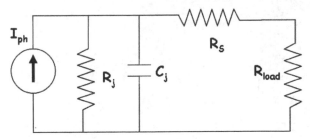

Fig. 6. Small-signal circuit associated to IPE-based device.

The RC limited bandwidth is given by:

$$f_{RC} = \frac{1}{2\pi\left[(R_s + R_{load})//R_j\right]C_j} = \frac{1}{2\pi R_{Tot}C_j} \qquad (17)$$

where R_j can be evaluated from the inverse derivative of the reverse current-voltage (I-V) electrical characteristic (tipical values are in the MΩ range) and R_s can be extracted from the forward I-V characteristic.

For high speed applications, the load resistance R_L, tipically 50 Ω, is much lower than series resistance (R_s) and junction resistance (R_j). Therefore, the device 3-dB frequency, becomes (Donati, 1999):

$$f_{3dB} = \frac{1}{2\pi R_s C_j} \qquad (18)$$

Where the junction capacitor is linked to the photodetector area (A_{Ph}) from the following formula:

$$C_j = \frac{\varepsilon_{Si} A_{Ph}}{W} \qquad (19)$$

with ε_{Si} is the silicon dielectric constant (10^{-12} C/cmV). It is worth noting that when the detector area is made sufficiently small, the influence due to the capacitance is reduced, the effect of the transit time dominates and high speed operation can be reached.

The intrinsic carrier-transit time limited 3-dB bandwidth for the device is given by (Donati, 1999):

$$f_{tr} = \frac{0.44}{t_d} = 0.44\frac{v_t}{L} \qquad (20)$$

where v_t is the effective carrier saturation velocity (10^7cm/s in Si) and L is the carrier transit distance. In Fig. 7, the frequency due to RC time constant and transit time are reported as a function of the detector area for R_L= 50 Ω an W=1 μm. It is worth noting that when the detector area is made sufficiently small, i.e. smaller than $A_{Ph}<65\mu m^2$, the influence due to the capacitance is reduced of one order to magnitude with respect to the transit time, enabling device to work at GHz range.

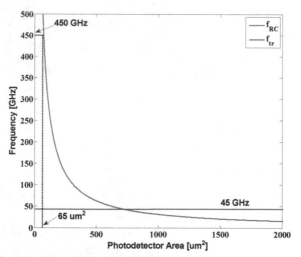

Fig. 7. Frequency due to the RC time and the transit time plotted against area detector for
RL=50 Ω and W=1μm.

2.3 Noise and sensitivity of IPE-based devices

It is well known that the photodetector output is affected by a noise contribution which
hinders the device sensitivity. The r.m.s. noise is due to two contributions: Johnson noise
and shot noise. Johnson is the thermal noise associated to the resistance R_{load} and its r.m.s
value can be defined as (Donati, 1999):

$$i_J = \sqrt{\frac{4kBT}{R_{load}}} \tag{21}$$

where R_{load} is the load resistance, k is the Boltzmann constant and T is the absolute
temperature. On the other hand, the shot noise is associated to the discrete nature of the
total current, i.e., the sum of the signal current and dark current (Donati, 1999):

$$i_S = \sqrt{2q(I_{ph}+I_d)B} \tag{22}$$

where q is the electron charge.

Because the two contributions are statistically independent their m.s. values can be added in
order to get the total noise:

$$i_n = \sqrt{i_S^2 + i_J^2} = \sqrt{2q(I_{ph}+I_d)B + \frac{4kBT}{R}} \tag{23}$$

From Eq. 23, it should be clear that shot noise can always dominate the Johnson with a right
choice of R_{load}. Dark current of infrared photodetectors based on the IPE effect is composed by
the inverse saturation current (I_s) and by a background current (I_{bg}). The background current is
due to the fact that at first approximation an infrared photodetector at a temperature T, can be
view as a blackbody emitting a radiance r(λ) according to the Planck law (Donati, 1999):

$$r(\lambda) = \frac{2h\nu^2}{\lambda^3} \left[\frac{1}{e^{\frac{h\nu}{kT}} - 1} \right] \tag{24}$$

where hν and λ are the photon energy and wavelength, respectively. It could be shown that this radiance is proportional to a dc background current (I_{bg}) limiting infrared photodetector sensitivity. In order to get higher sensitivity, is necessary to reduce the background current contribute by reducing the emission radiance, i.e., by lowering the photodetector temperature (see Plank law). For this reason infrared detector needs to work at cryogenic temperature. On the other hand, photodetectors working at visible or NIR wavelengths are not able to detect blackbody emission and the Schottky junction saturation current limits the detector sensitivity. It is given by the Richardson-Dushamann equation (Yuan & Perera, 1995):

$$I_s = A_{ph} A^{**} T^2 e^{-\frac{\phi_B(V)}{kT}} \tag{25}$$

where $A^{**} = f_p A^*$, A^* is the Richardson constant (30 A/cm²K² for p-type Si and 110 A/cm²K² for n-type Si) (Sze, 1981), f_p is the barrier escape probability which as a first approximation is given by $f_p(V) = \exp(-x_m(V)/L_s)$, L_s is the electron scattering length, T is the absolute temperature and $\Phi_B(V) = \Phi_{B0} - \Delta\Phi_B(V)$ is the potential barrier which is the potential barrier at zero voltage minus the lowering due to the reverse bias. It is worth noting that both potential barrier and barrier escape probability are reverse voltage dependent due to force image effect. The plot of the saturation current density against reverse voltage applied for three different metal: gold (Au), silver (Ag) and copper (Cu), is reported in Fig. 8 (Casalino et al., 2008b). Schottky barrier used in the simulations for Au/p-Si, Ag/p-Si and Cu/p-Si interfaces are 0.78 eV, 0.78 eV and 0.58 eV, respectively. Due to the lowest potential barrier, copper shows the highest dark current density.

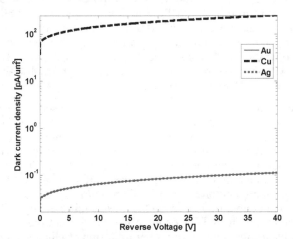

Fig. 8. Dark current density plotted against reverse voltage (semi-log scale) for three different metals: gold, silver and copper. Reprinted with permission from M. Casalino et al., "A silicon compatible resonant cavity enhanced photodetector working at 1.55 μm," Semicond. Sci. Technol., 23, 075001, 2008 (doi: 10.1088/0268-1242/23/7/075001). IOP Publishing is acknowledged.

The photodetector sensitivity is directly linked to the r.m.s. noise current and can be represented in term of a well known parameter called NEP (Noise Equivalent Power) which is defined as the ratio of r.m.s. noise to responsivity (Donati, 1999):

$$NEP = \frac{i_n}{\Re} \ [W] \tag{26}$$

NEP represents the lowest input power giving a unit signal/noise ratio. Because higher device performances correspond to smaller NEP, it is more convenient to define its inverse: the detectivity. It can be expected that the square of the r.m.s. noise (m.s. noise) is proportional to the electrical bandwidth (B) and detector area (A_{Ph}), for this reason in order to get a figure of merit not depending on these parameters, the detectivity is normalized to the square root of detector area and bandwidth (Donati, 1999):

$$D = \sqrt{A_{Ph} \cdot B} \frac{\Re}{i_n} \ \left[\frac{cm\sqrt{Hz}}{W} \right] \tag{27}$$

3. Silicon photodetectors based on IPE

This section reviews the history and the progresses of the main structures reported in literature, starting from historical devices for imaging application at infrared wavelengths, up to new NIR devices adapted for telecommunications and power monitoring applications.

3.1 Infrared devices for imaging application

Schottky-barrier focal plane arrays (FPAs) are infrared imagers, fabricated by well-established silicon very-large-scale-integration (VLSI) process, representing one of the most effective technology for large-area high-density focal plane arrays for many near infrared (NIR, 1 to 3 μm) and medium infrared (MIR, 3 to 5 μm) applications. PtSi Schottky Barrier detectors (SBDs) represent the most established SBD technology but they must operate at 77 °K in order to reduce the dark current density in the range of a few nA/cm². Pd$_2$Si SBDs were developed for operation with passive cooling at 120 °K in the NIR band, while IrSi SBDs have also been investigated to extend the application of Schottky-barrier focal plane arrays into the far infrared (FIR, 8 to 10 μm) spectral range.

In 1973 Shepherd and Yang (Shepherd & Yang, 1973) proposed the concept of silicide Schottky-barrier detectors as much more reproducible alternative to HgCdTe FPAs for infrared thermal imaging. After being dormant for about ten years, extrinsic Si was reconsidered as a material for infrared imaging, especially after the invention of charge-coupled devices (CCDs) by Boyle and Smith (Boyle & Smith, 1970). For the first time it became possible to have much more sophisticated readout schemes and both detection and readout could be implemented on one common silicon chip. Since then, the development of the Schottky-barrier technology progressed continuously and currently offers large IR image sensor formats. These trends in IR FPA development show that the IR community today prefers more producible technologies with higher uniformity to the technology based on the narrow gap semiconductors, which still have serious material problems. Such attributes as: monolithic construction, uniformity in responsivity and signal to noise (the performance of an IR system ultimately depends on the ability to compensate the non uniformity of an FPA using external electronics and a variety of temperature references), and absence of discernible 1/f noise; make Schottky-barrier devices a formidable contender to the mainstream infrared systems and applications. (Shepherd, 1984; Shepherd, 1988; Kosonocky, 1991; Kosonocky, 1992; Shepherd, 1998). In the early years, the development of SBD FPA technology progressed from the demonstration of the initial concepts in the 1970's (Kohn et al., 1975; Capone et al., 1978; Kosonocky et al., 1978; Shepherd et al., 1979) to the development of high resolution scanning and staring devices in the 1980's and in the 1990's, that were at the basis of many applications for infrared imaging in the NIR, MIR and FIR bands. The first Schottky-barrier FPAs were made with thick Pd2Si or PtSi detectors using about 600 Å of deposited palladium or platinum. These FPAs exhibited relatively small photoresponse. More than an order of magnitude improvement in photoresponse was demonstrated in 1980 with 50×50-element FPAs constructed with thin PtSi SBDs at the David Sarnoff Research Center (Taylor et al, 1980; Kosonocky et al., 1980). The SBDs in this FPA had an optical cavity in the form of a thin (20 – 40 Å) PtSi layer separated from an aluminium reflector by a layer of deposited SiO2. The general concept of thin SBD with optical cavity was first described in 1973 by Archer and Cohen for SBDs in the form of Au on p-type Si (Archer & Cohen, 1973). The improved PtSi SBD structure was further developed and from 1980 to 1985 the fabrication process for PtSi SBDs was optimized at Sarnoff with the development of 32×63, 64×128 and 160×244 IR-CCD FPAs (Elabd et al., 1982; Kosonocky, 1985). Similar PtSi SBDs characteristics were also reported in the same years by Mitsubishi Corporation, Fujitsu, NEC, EG&G Reticon, Hughes, Loral Fairchild and Kodak.

As already discussed, the most popular Schottky-barrier detector is the PtSi detector, which can be used for the detection in the 3-5 μm spectral range (Kimata & Tsubouchi, 1995; Kimata, 2000). Radiation is transmitted through the p-type silicon and is absorbed in the metal PtSi (not in the semiconductor), producing hot holes, which are then emitted over the potential barrier into the silicon, leaving the silicide charged negatively. Negative charge of silicide is transferred to a CCD by the direct charge injection method (see Fig. 9).

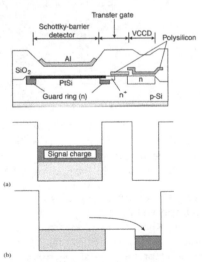

Fig. 9. Typical construction and operation of PtSi Schottky-barrier IR FPA designed with interline transfer CCD readout architecture. (a) and (b) show the potential diagrams in the integration and readout operations, respectively (Kimata, 1995, 1998). Reprinted with permission from M. Kimata, "PtSi Schottky-barrier infrared focal plane arrays," Opto-Electronics Review, 6, 1, 1998.

The effective quantum efficiency in the 3–5 μm window is very low, of the order of 1%, but useful sensitivity is obtained by means of near full frame integration in area arrays. The quantum efficiency has been improved by thinning PtSi film and implementation of an optical cavity. Due to very low quantum efficiency, the operating temperature of Schottky-barrier photoemissive detectors is lower than another types of IR photon detectors (Rogalsky, 1999).

Schottky photoemission is independent of such factors as semiconductor doping, minority carrier lifetime, and alloy composition, and, as a result of this, has spatial uniformity characteristics that are far superior to those of other detector technologies. Uniformity is only limited by the geometric definition of the detectors. The fundamental source of dark current in the devices is thermionic emission of holes over the potential barrier and its magnitude is given by Richardson's equation (Sze, 1982). The cooling requirements of photoemissive detectors are comparable to the extrinsic devices, and while an extension of the technology to the long wavelength band is possible using IrSi (see Fig. 10) this will require cooling below 77 °K (Shepherd, 1988).

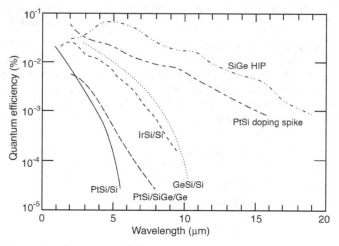

Fig. 10. Comparison of various infrared detectors based on IPE for PtSi, IrSi, PtSi/SiGe, PtSi doping spike, and SiGe/Si HIP (Kimata, 2000). Reprinted with permission from M. Kimata, Metal silicide Schottky infrared detector arrays. In: Infrared Detectors and Emitters: Materials and Devices, Kluwer Academic Publishers, Boston, USA, 2000.

The Schottky-barrier detector is typically operated in backside illumination mode. The quantum efficiency has been improved by thinning PtSi film. The thinning is effective down to the PtSi thickness of 2 nm. Another means of improving responsivity is implementation of an optical cavity. The optical cavity structure consists of the metal reflector and the dielectric film between the reflector and metal electrode of the Schottky-barrier diode. According to fundamental optical theory, the effect of the optical cavity depends on the thickness and refractive index of the dielectric films, and the wavelength (Kurianski et al., 1989). The main advantage of the Schottky-barrier detectors is that they can be fabricated as monolithic arrays in a standard silicon VLSI process. Typically, the silicon array is completed up to the Al metallization al step. A Schottky-contact mask is used to open SiO_2

surface to p-type <100> silicon (with resisitivity 30–50 Ω/cm) at the Schottky-barrier detector location. In the case of PtSi detectors, a very thin layer of Pt (1–2 nm) is deposited and sintered (annealed at a temperature in the range of 300–600°C) to form PtSi and the unreacted Pt on the SiO_2 surfaces is removed by dip etching in hot aqua regia. The Schottky-barrier structure is then completed by a deposition of a suitable dielectric (usually SiO_2) for forming the resonant cavity, removing this dielectric outside the Schottky-barrier regions, and depositing and defining Al for the detector reflector and the interconnects of the Si readout multiplexer. In the case of the 10 μm IrSi Schottky-barrier detectors, the IrSi was formed by in situ vacuum annealing and the unreacted Ir was removed by reactive ion etching. The progress of the Schottky-barrier FPA technology has been constant (Kimata et al., 1998). At the present time Schottky-barrier FPAs represent the most advanced FPAs technology for medium wavelength applications (see Table 1).

Array size	Readout	Pixel size (μm^2)	Fill factor (%)	Saturation (e^-)	NEDT/(f/#) (K)	Year	Company
512 × 512	CSD	26 × 20	39	1.3×10^6	0.07(1.2)	1987	Mitsubishi
512 × 488	IL-CCD	31.5 × 25	36	5.5×10^5	0.07(1.8)	1989	Fairchild
512 × 512	LACA	30 × 30	54	4.0×10^5	0.10(1.8)	1989	RADC
640 × 486	IL-CCD	25 × 25	54	5.5×10^5	0.10(2.8)	1990	Kodak
640 × 480	MOS	24 × 24	38	1.5×10^6	0.06(1.0)	1990	Sarnoff
640 × 488	IL-CCD	21 × 21	40	5.0×10^5	0.10(1.0)	1991	NEC
640 × 480	HB/MOS	20 × 20	80	7.5×10^5	0.10(2.0)	1991	Hughes
1040 × 1040	CSD	17 × 17	53	1.6×10^6	0.10(1.2)	1991	Mitsubishi
512 × 512	CSD	26 × 20	71	2.9×10^6	0.03(1.2)	1992	Mitsubishi
656 × 492	IL-CCD	26.5 × 26.5	46	8.0×10^5	0.06(1.8)	1993	Fairchild
811 × 508	IL-CCD	18 × 21	38	7.5×10^5	0.06(1.2)	1996	Nikon
801 × 512	CSD	17 × 20	61	2.1×10^6	0.04(1.2)	1997	Mitsubishi
1968 × 1968	IL-CCD	30 × 30	—	—	—	1998	Fairchild

Table 1. Specifications and performances of typical PtSi Schottky-barrier FPAs (Kimata, 2000). Reprinted with permission from M. Kimata, Metal silicide Schottky infrared detector arrays. In: Infrared Detectors and Emitters: Materials and Devices, Kluwer Academic Publishers, Boston, USA, 2000.

The details of the geometry, and the method of charge transfer differ for different manufacturers. The design of a staring Schottky-barrier FPAs for given pixel size and design rules, involves a trade-off between the charge handling capacity and the fill factor. Most of the reported Schottky-barrier FPAs have the interline transfer CCD architecture. The typical cross section view of the pixel and its operation in interline transfer CCD architecture is shown in Fig. 9. The pixel consists of a Schottky-barrier detector with an optical cavity, a transfer gate, and a stage of vertical CCD. The n-type guard ring on the periphery of the Schottky-barrier diode reduces the edge electric field and suppresses dark current. The effective detector area is determined by the inner edge of the guard ring. The transfer gate is an enhancement MOS transistor. The connection between detector and the transfer gate is made by an n+ diffusion. A buried-channel CCD is used for the vertical transfer. During the optical integration time the surface-channel transfer gate is biased into accumulation. The Schottky-barrier detector is isolated from the CCD register in this condition. The IR radiation generates hot holes in the PtSi film and some of the excited hot holes are emitted into the silicon substrate leaving excess electrons in the PtSi electrode. This lowers the electrical potential of the PtSi electrode. At the end of the integration time, the transfer gate is pulsed-on to read out the signal electrons from the detector to the CCD register. At the same time, the electrical potential of the PtSi electrode is reset to the channel level of the transfer gate. A unique feature of the Schottky-barrier IR FPAs is the built-in blooming control (blooming is a form of crosstalk in which a well saturates and the electrons spill over into neighbouring pixels). A strong illumination forward biases the detector and no further electrons are accumulated at the detector. The small negative voltage developed at the detector is not sufficient to forward bias the guard ring to the extent that electrons are injected to the CCD register through the silicon region under the transfer gate. Therefore, unless the vertical CCD has an insufficient charge handling capacity, blooming is suppressed perfectly in the Schottky-barrier IR FPA. The responsivity of the FPAs is proportional to their fill factor, and improvement in the fill factor has been one of the most important issues in the development of imagers. For improving the fill factor a readout architecture called the charge sweep device (CSD) developed by Mitsubishi Corporation is also used. Kimata and co-workers have developed a series of IR image sensors with the CSD readout architecture with array sizes from 256×256 to 1040×1040 elements. Specifications and performance of these devices are summarised in Table 2. The effectiveness of this readout architecture is enhanced as the design rule becomes finer. Using a 1.2 μm CSD technology, a large fill factor of 71% was achieved with a 26×20 μm^2 pixel in the 512×512 monolithic structure (Yagi et al., 1994). The noise equivalent temperature difference (NETD) was estimated as 0.033 °K at 300 °K. The 1040×1040 element CSD FPA has the smallest pixel size (17×17 μm^2) among two-dimensional IR FPAs. The pixel was constructed with 1.5 μm

Array size	256 × 256	512 × 512	512 × 512	512 × 512	801 × 512	1040 × 1040
Pixel size (μm^2)	26 × 26	26 × 20	26 × 20	26 × 20	17 × 20	17 × 17
Fill factor (%)	58	39	58	71	61	53
Chip size (mm^2)	9.9 × 8.3	16 × 12	16 × 12	16 × 12	16 × 12	20.6 × 19.4
Pixel capacitor	Normal	Normal	High-C	High-C	High-C	High-C
CSD	4-phase	4-phase	4-phase	4-phase	4-phase	4-phase
HCCD	4-phase	4-phase	4-phase	4-phase	4-phase	4-phase
Number of outputs	1	1	1	1	1	4
Interface	Non integration	Field integration	Frame/Field integration	Frame/Field integration	Flexible	Field integration
Number of I/O pins	30	30	30	30	25	40
Process technology	NMOS/ CCD	NMOS/ CCD	NMOS/ CCD	NMOS/ CCD	CMOS/ CCD	NMOS/ CCD
	2 poly/2 Al	2 poly/2 Al	2 poly/2 Al	2 poly/2 Al	2 poly/2 Al	2 poly/2 Al
Design rule (μm)	1.5	2	1.5	1.2	1.2	1.5
Thermal response (ke/K)	—	13	—	32	22	9.6
Saturation (e)	0.7 × 10^6	1.2 × 10^6	—	2.9 × 10^6	2.1 × 10^6	1.6 × 10^6
NETD (K)	—	0.07	—	0.033	0.037	0.1

Table 2. Specifications and performance of 2-D PtSi Schottky-barrier FPAs with CSD readout (Kimata, 1998). Reprinted with permission from M. Kimata, "PtSi Schottky-barrier infrared focal plane arrays," Opto-Electronics Review, 6, 1, 1998.

design rules and has 53% fill factor (Kimata et al., 1992). If the signal charges of 1040×1040 pixels are readout from one output port at the TV compatible frame rate, an unrealistic pixel rate of about 40 MHz is required. Therefore, a 4-output chip design was adopted (Shiraishi et al., 1996). The array of 1040×1040 pixels is divided into four blocks of 520×520 pixels. Each block has a horizontal CCD and a floating diffusion amplifier. One-million pixel data at a 30 Hz frame rate can be readout by operating each horizontal CCD at a 10 MHz clock frequency. The NEDT of 1040×1040 element FPA at 300 °K with a 30 Hz frame is 0.1 °K. More recently, a high-performance 801×512-element PtSi Schottky-barrier infrared image sensor has been developed with an enhanced CSD readout architecture (Inoue et al., 1997; Kimata et al., 1997). The developed image sensor has a large fill factor of 61% in spite of a small pixel size of 17×20 μm². The NEDT was 0.037 °K at 300 °K. The total power consumption of the device was less than 50 mW. Current PtSi Schottky barrier FPAs are mainly manufactured in 150 mm wafer process lines with around 1 μm lithography technologies; the most advanced Si technology offers 200 mm wafers process with 0.25 μm design rules. Furthermore, 300 mm Si wafer processes with 0.15 μm fine patterns will soon available. However, the performance of monolithic PtSi Schottky-barrier FPAs has reached a plateau, and a slow progress from now on is expected.

3.2 IPE-based NIR devices

In this section, the main Near-Infrared silicon photodetectors structures based on the afore-mentioned IPE effect will be reviewed. The simplest detectors consist of a metal layer on a semiconductor forming a Schottky contact at the material interface, with the Schottky barrier energy Φ_B determined by the materials (Sze, 1981). If an infrared radiation reaches metal/semiconductor interface, the conduction electrons inside the metal can absorb photons gaining sufficient energy. These excited (hot) electrons are able to cross over the Schottky barrier, sweep out the depletion region of the semiconductor, and be collected as a photocurrent under reverse bias operation from an ohmic semiconductor-metal interface. In practice, p-type silicon is often used as the semiconductor because Schottky barriers are lower thereupon than on n-type silicon, allowing detection at longer wavelengths. The IPE-effect, as described in the section 2.1, is very fast allowing to reach high data rates. On the other hand, it is inherently weak so several photodetector designs have been proposed for achieving a suitable quantum efficiency. Elabd et al. proposed to use a very thin metal film (about 2 nm) to increase the escape probability of hot carriers due to their multiple reflections inside the metal film (Elabd et al., 1982). In particular, the proposed Schottky photodetector was characterized by a responsivity of about 250 mA/W for a wavelength λ=1500 nm. However, since the volume in which the photons interact with electrons in the metal is very small, only a small fraction of the incident photons actually causes photoemission. Several solutions have been proposed to enhance the efficiency of the IPE process. For example, in the 2001 Wang's group (Lee et al., 2001) investigated the spectral responsivity of Al-porous silicon Schottky barrier photodetectors in the wavelength range 0.4-1.7 μm. The structure of the PS photodetector was Al (finger type)/PS/Si/Al (ohmic), and the active area was 18 mm². The photodetectors show strong photoresponsivity in both the visible and the infrared bands, especially at 1.55 μm. The photocurrent can reach 1.8 mA at a reverse bias of 6 V under illumination by a 1.55-μm, 10-mW laser diode. The corresponding quantum efficiency is 14.4%; this high value comes from a very high surface-area-to-volume ratio, of the order of 200-800 m²/cm³ of porous Si. The dark current is ~5 μA at -10 V. Recently, Casalino *et al.* (Casalino et al., 2008a, 2010b) propose to enhance the IPE absorption by a resonant cavity effect. The conceptual scheme of the proposed device in shown in Fig. 11.

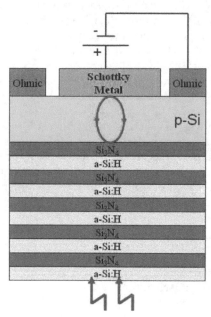

Fig. 11. Schematic cross section of the photodetector proposed by Casalino *et al.* (Casalino et al., 2010b) in 2010.

The resonant cavity is a vertical-to-the-surface Fabry-Perot structure. It is formed by a buried reflector, a top mirror interface and, in the middle, a silicon cavity. The buried reflector is a Bragg mirror, realized by alternating layers of amorphous hydrogenated silicon (a–Si:H) and silicon nitride (Si_3N_4). A Schottky metal layer (Cu), working both as active (absorbing) layer and as cavity mirror, is deposited above the silicon layer. The room temperature responsivity measurements on the device return a peak value of about 2.3 µA/W and 4.3 µA/W, respectively for 0 V and -10 mV of reverse bias applied (Casalino et al., 2008a). In (Casalino et al., 2010b) the authors propose both to scaling down and optimize the same device described in (Casalino et al., 2008a). In particular, the cavity finesse was increased allowing to reach a responsivity of 8 µA/W for a reverse bias of 100mV.

A substantial enhancement of the IPE process efficiency has been achieved increasing the interaction of light with the metal in the vicinity of the interface by the confinement of the infrared radiation into a silicon waveguide. This solution has effectively proven that Schottky diode photodetectors are good candidate for the highly integrated photonics circuits. An example of this approach has recently demonstrated by Coppola's group (Casalino et al., 2010c). The proposed device is schematically illustrated in Fig. 12. A rib waveguide was terminated on a deep trench that reaches down the buried oxide layer of the SOI wafer. A Cu/p-Si Schottky contact was fabricated on the vertical surface of the deep trench.

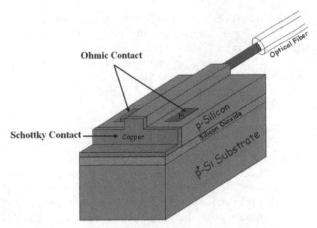

Fig. 12. Schematic view of the Cu/p-Si Schottky barrier-based integrated photodetector proposed by Casalino et al. (Casalino et al., 2010c). Reprinted with permission from M. Casalino *et al.*, "Cu/p-Si Schottky barrier-based near infrared photodetector integrated with a silicon-on-insulator waveguide," Appl. Phys. Lett., 96, 241112. Copyright 2010 American Institute of Physics.

By means of this technological solution a very narrow semiconductor/metal barrier transverse to the optical field coming out from the waveguide has been achieved. The integrated photodetector was characterized by a responsivity of 0.08 mA/W at a wavelength of 1550 nm with an reverse bias of -1V. Measured dark current at -1 V was about 10 nA. Moreover, the authors assert that the thinness of Cu/p-Si Schottky barrier could enable a speed operation in the gigahertz range. An indirect evaluation of the bandwidth of the detector was reported to confirm the operation speed potentialities. A bandwidth of about 3 GHz was measured by Zhu *et al.* in (Zhu et al., 2008a) on a Schottky barrier based integrated photodetector, where the junction was achieved by a nickel silicide layer (NiSi$_2$) on silicon (Zhu et al., 2008b, 2008c)(Fig. 13).

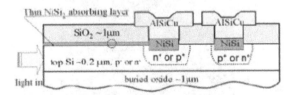

Fig. 13. Schematic structure of waveguide-based silicide Schottky-barrier photodetector proposed by Zhu et al. (Zhu et al., 2008b, 2008a). Reprinted with permission from S. Zhu *et al.*, "Near-infrared waveguide-based nickel silicide Schottky-barrier photodetector for optical communications," Appl. Phys. Lett., 92, 081103. Copyright 2008 American Institute of Physics.

In particular, the authors proposed the lengthening of a thin silicide layer on the surface of a SOI waveguide to achieve both a suitable optical absorption and an efficient photoexcitation of metal electrons or holes across the silicide/Si interface. A detailed analysis was reported on the influence of the silicide layer dimension on the performances of both NiSi$_2$/p-Si and NiSi$_2$/n-Si diodes. The overall performances of the NiSi$_2$/p-Si structure have been resulted better than that relative to the NiSi$_2$/n-Si interface, due to the lower Schottky barrier height of the latter structure. In particular, a responsivity of about 4.6 mA/W at a wavelength of 1550nm and a reverse bias of -1V was estimated for the NiSi$_2$/p-Si diode against the value of 2.3 mA/W of the NiSi$_2$/n-Si junction. Moreover, also the 3 nA of the measured dark current can be considered acceptable.

In order to increase further the capability of IPE-based silicon photodetector to detect long-wavelength (infrared) photons, the possibility to use the concept of surface plasmon polaritons (SPPs) has been explored. SPP are TM-polarized electromagnetic waves trapped at or guided along a metal-dielectric interface. This effect was discovered several decades ago but attention has been renewed by the phenomena of enhanced optical transmission through metallic films with nanostructure (Reather, 1988; Ebbesen et al., 1998). In fact, SPPs are shorter in wavelength than the incident light providing a significant increase in spatial confinement and local field intensity. In other words, while optical systems are basically diffraction limited, surface plasmon polaritons allow a tight localization of optical field to strongly subwavelength dimensions at the metal-dielectric interface (Barnes et al., 2003; Maier, 2006). Such property means that the infrared light is directly guided toward the active area of the IPE detector, confining the optical power at the boundary between the materials forming the Schottky contact, thereby increasing the interaction of light with the metal in the vicinity of the interface where the photoemission process takes place. Berini's group at University of Ottawa has recently applied these concepts confirming that the high light confinement of the SPPs structures can significantly improve the detection capability of the IPE-based photodetector integrated on Si wafers. In particular, Scales *et al.* in (Scales et al., 2004, 2009) described a Schottky diode photodetector obtained embedding a metal stripe of finite width in a homogeneous dielectric cladding (symmetric structure). The proposed device (see Fig. 14) was characterized by a responsivity of about 0.1 A/W, a dark currents of 21 nA, and minimum detectable powers of -22 dBm at a wavelength λ=1550 nm.

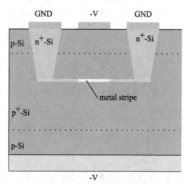

Fig. 14. SPPs Schottky detector based on a metal stripe surrounded by silicon.

An accurate investigation of the performance of the same device for different wavelengths and for several metals forming Schottky contacts were carried out in (Scales et al., 2011). It is worth noting that in these symmetric structures the realization of a thin metal film buried in a semiconductor become the fabrication process more complicated, however, despite of this drawback, the IPE is enhanced due to emission carriers occurring through two Schottky barriers (Scales & Berini, 2010). In order to simplify the fabrication process and to obtain a very short device, Berini's group has proposed an asymmetric SPPs-based photodetector (Akbari et al., 2009, 2010). The proposed structure supports highly confined and highly attenuated SPP modes; this latter feature allows to fabricate shorter devices compared to the symmetric photodetectors. The device, sketched in Fig. 15, consists of a metal stripe cladded at the bottom by a layer of Silicon and covered by air and exhibits for a wavelength λ=1280nm a maximum value of responsivity of about 1 mA/W with a dark current of 6 µA (Akbari et al., 2010).

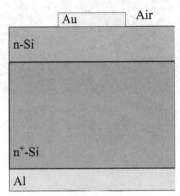

Fig. 15. Cross-section of the aymmetric SPPs Schottky detector proposed in (Akbari et al., 2010).

A detailed simulation analysis performed on asymmetric Schottky detector shows that a significant enhancement in the responsivity can be achieved for a thin metal stripe (about 5 nm) due to multiple internal reflections of excited carriers (Akbari et al., 2009). In particular, authors report that the enhancement is more noticeable for thin metal stripe on p-Si compared to device on n-Si. In this case, the energy range over which the hot electrons can experience multiple reflections is very small because of the larger Schottky barrier height. Finally, the same group has recently demonstrated that a considerable increase in responsivity can be reached with strong applied reverse bias (Olivieri et al., 2010). In particular, authors driving a no optimized detector like that shown in Fig. 15 into breakdown (V~–210 V), such that internal electronic gain is obtained by carrier multiplication, obtaining a responsivity of 2.35 mA/W.

As mentioned above the possibility to achieve a strong light confinement implies both an improvement of responsivity and an important advancement in device miniaturization enabling the realization of on-chip photodetectors on the nanoscale. In this contest, Goykhman et al. in (Goykhmanet al., 2011) have been characterized a nanoscale silicon surface-plasmon Schottky detector shown in Fig. 16.

The detector was fabricated employing a self-aligned approach of local-oxidation of silicon (LOCOS) on silicon on insulator substrate. This approach has been proved useful for fabricating in the same process both a low-loss bus photonic waveguide and the detector. Actually, the oxide spacers effectively define the nanometric area of metal-silicon interface and thus allow avoiding lateral misalignment between the silicon surface and the metal layer to form a Schottky contact. The so realized photodetector was characterized by a responsivity 0.25, 1.4, and 13.3 mA/W for incident optical wavelengths of 1.55, 1.47, and 1.31 μm, respectively.

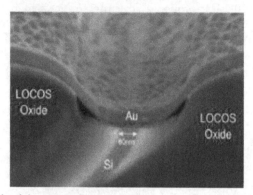

Fig. 16. SEM micrograph of the nanoscale Schottky contact proposed in (Goykhman et al., 2011). Reprinted with permission from I. Goykhman et al., "Locally Oxidized Silicon Surface-Plasmon Schottky Detector for Telecom Regime," Nano Lett. 11, 2219–2224. Copyright 2011 american Chemical Society.

4. Conclusion

In this chapter an overview on the NIR all-Si photodetectors based on the IPE has been presented. First, we have attempted to elucidate the IPE effects allowing Si absorption at sub band-gap wavelengths and the main figures of merit of IPE-based devices. Then, a quantitative comparison of the photodetectors proposed in the scientific literature, including both bulk and integrated devices, have been reviewed. Unfortunately, the efficiency of devices is low compared to that of detectors based on inter-band absorption. This property is a direct result of many factors: 1) the low absorption due to high reflectivity of the metal layer at NIR wavelengths, 2) the conservation of momentum during carrier emission over the potential barrier which lowers the carriers emission probability into semiconductor, 3) the excitation of carriers lying in states far below the Fermi energy, which get very low probability to overcome the Schottky barrier. While the first and second points can be partially improved by using an optical cavity and a thin metal film, respectively, the third point remains the main limiting factor of IPE. In 1971 Sheperd, Vickers and Yang (Sheperd et al., 1971) recommended replacing the metal electrodes IPE-based photodiodes with degenerate semiconductors. They reasoned that reducing the Fermi energy of electrode by substitution of a degenerate semiconductor could improve emission efficiency as much as 20-fold. Even if NIR IPE-based devices efficiency is still only adapted for power monitoring application, in our opinion that new structures based on the aforementioned insights could play a key role in telecommunication field and could open new frontiers in the field of low-cost silicon photonic.

5. References

Akbari A. & Berini, P. (2009). Schottky contact surface-plasmon detector integrated with an asymmetric metal stripe waveguide. *Appl. Phys. Lett.*, Vol.95, No.2, pp. 021104.

Akbari, A.; Tait, R. N. & Berini, P. (2010). Surface plasmon waveguide Schottky detector. *Opt. Express*, Vol.18, No.8, pp. 8505- 8514.

Archer, R.J. & Cohen, J. (1973). Schottky-Barrier Monolithic Detector Having Ultrathin Metal Layer. *U.S. Patent 3,757,123*

Barnes, W. L.; Dereux, A. & Ebbesen, T.W. (2003) Surface plasmon subwavelength optics. *Nature*, Vol.424, No.6950, pp. 824–830

Boyle, W.S. & Smith, G.E. (1970). Charge-coupled semiconductor devices. *Bell Syst. Tech. J.*, Vol.49, pp. 587–593

Bradley, J.D.B.; Jessop, P.E. & Knights, A.P. (2005). Silicon waveguide-integrated optical power monitor with enhanced sensitivity at 1550 nm. *Appl. Phys. Lett.*, Vol. 86, No.24, pp. 241103.

Capone, B.R.; Skolnik, L.H.; Taylor, R.W.; Shepherd, F.D.; Roosild, S.A.; Ewing, W.; Kosonocky, W.F. & Kohn, E.S. (1978). Evaluation of Schottky IRCCD Staring Mosaic Focal Plane. *22nd Int. Tech. Symp. Society of Photo-Optical Instrumentation Engineers (SPIE)*, San Diego, Aug. 28-29

Casalino, M. ; Sirleto, L. ; Moretti, L. ; Della Corte, F. & Rendina, I. (2006a). Design of a silicon RCE schottky photodetector working at 1.55 micron. *Journal of luminescence*, Vol.121, pp. 399-402.

Casalino, M. ; Sirleto, L. ; Moretti, L. ; Della Corte, F. & Rendina, I. (2006b). Design of a silicon resonant cavity enhanced photodetector based on the internal photoemission effect at 1.55 µm. *Journal of Optics A: Pure and applied optics*, Vol.8, pp. 909-913.

Casalino, M.; Sirleto, L.; Moretti, L.; Gioffrè, M.; Coppola, G. & Rendina, I. (2008a). Silicon resonant cavity enhanced photodetector based on the internal photoemission effect at 1.55 µm: Fabrication and characterization. *Appl. Phys. Lett.*, Vol.92, No.25, pp. 251104

Casalino, M.; Sirleto, L.; Moretti, L. & Rendina, I. (2008b). A silicon compatible resonant cavity enhanced photodetector working at 1.55 µm. *Semicond. Sci. Technol.*, Vol.23, No.7, pp. 075001

Casalino, M. ; Coppola, G. ; Iodice, M. ; Rendina, I. & Sirleto, L. (2010a). Near Infrared All-Silicon Photodetectors: State of the Art and Perspectives. *Sensors*, Vol.10, No.12, pp. 10571-10600

Casalino, M.; Coppola, G.; Gioffrè, M.; Iodice, M.; Moretti, L.; Rendina, I. & Sirleto, L. (2010b). Cavity enhanced internal photoemission effect in silicon photodiode for sub-bandgap detection. *J. Lightw. Technol.*, Vol. 28, No.22, pp. 3266-3272

Casalino, M.; Sirleto, L.; Iodice, M.; Saffioti, N.; Gioffrè, M.; Rendina, I & Coppola, G. (2010c). Cu/p-Si Schottky barrier-based near infrared photodetector integrated with a silicon-on-insulator waveguide. *Appl. Phys. Lett.*, Vol.96, No.24, pp. 241112.

Chan, E.Y; Card, H.C. & Teich, M.C. (1980a). Internal Photoemission Mechanism at interfaces between Germanium and Thin Metal Films. *IEEE Journal of Quantum Electronics*, Vol.QE-16, No.3, pp. 373-381

Chan, E.Y. & Card, H.C. (1980b). Near IR interband transitions and optical parameters of metal-germanium contacts. *Applied Optics*, Vol.19, No.8, pp. 1309

Donati, S. (1999). *Photodetectors: Devices, circuits, and applications*. Prentice Hall PTR, New Jersey, USA

Ebbesen, W. ; Lezec, H.J. ; Ghaemi, H.F. ; Thio, T. & Wolff, P.A. (1997). Extraordinary optical transmission through sub-wavelength hole arrays. *Nature, Vol.*391, pp 667-669

Elabd, H. & Kosonocky, W.F. (1982). Theory and measurements of photoresponse for thin film Pd$_2$Si and PtSi infrared Schottky-barrier detectors with optical cavity. *RCA Review*, Vol. 43, pp. 569–589

Elabd, H.; Villani, T. & Kosonocky, W.F. (1982). Palladium-Silicide Schottky-Barrier IR-CCD for SWIR Applications at Intermediate Temperatures. *IEEE Trans. Electron Devices Lett.*, Vol. EDL-3, pp. 89-90

Fowler, R.H. (1931). The Analysis of Photoelectric Sensitivity Curves for Clean Metals at Various Temperatures. *Physical Review*, Vol. 38, pp. 45-56

Goykhman, I.; Desiatov, B.; Khurgin, J.; Shappir, J. & Levy U. (2011). Locally Oxidized Silicon Surface-Plasmon Schottky Detector for Telecom Regime. Nano Lett. Vol. 11, pp. 2219–2224.

Inoue, M.; Seto, T.; Takahashi, S.; Itoh, S.; Yagi, H.; Siraishi, T.; Endo, K. & Kimata, M. (1997). Portable high performance camera with 801x512 PtSi-SB IRCSD, *SPIE Proc.*, Vol.3061, pp. 150–158

Jalali, B & Fathpour, S. (2006). Silicon Photonics. *J. Lightwave Technol.*, Vol.24, No.12, pp. 4600-4615

Kimata, M.; Yutani, N.; Tsubouchi, N. & Seto, T. (1992). High performance 1040x1040 element PtSi Schottky-barrier image sensor. *SPIE Proc.*, Vol.1762, pp. 350–360

Kimata, M. & Tsubouchi, N. (1995). Schottky barrier photoemissive detectors. In: *Infrared Photon Detectors*, A. Rogalski (Ed.), pp. 299-349, SPIE Optical Engineering Press, Bellingham

Kimata, M.; Ozeki, T.; Nunoshita, M. & Ito, S. (1997). PtSi Schottky-barrier infrared FPAs with CSD readout. *SPIE Proc.*, Vol.3179, pp. 212–223

Kimata, M.; Ueno, M.; Yagi, H.; Shiraishi, T.; Kawai, M.; Endo, K.; Kosasayama, Y.; Sone, T.; Ozeki, T. & Tsubouchi, N. (1998). PtSi Schottky-barrier infrared focal plane arrays, *Opto-Electronics Review*, Vol.6, pp. 1–10

Kimata, M. (2000). Metal silicide Schottky infrared detector arrays. In: *Infrared Detectors and Emitters: Materials and Devices*, Kluwer Academic Publishers, Boston, USA

Kimerling, L.C.; Dal Negro, L.; Saini, S.; Yi, Y.; Ahn, D.; Akiyama, S.; Cannon, D.; Liu, J.; Sandland, J.G.; Sparacin, D.; Michel, J.; Wada, K. & Watts, M.R. (2004). *Silicon Photonics: Topics in Applied Physics*, Springer, ISBN 3-642-05909-0, Berlin, Germania

Kohn, E.S.; Roosild, S.A.; Shepherd, F.D. & Yang, A.C. (1975). Infrared Imaging with Monolithic CDD-Addressed Schottky-Barrier Detector Arrays, Theoretical and Experimental Results. *Int. Conf. on Application on CCD's*, San Diego, Oct. 29-31

Kosonocky, W.F.; Kohn, E.S.; Shalleross, F.V.; Sauer, D.J.; Shepherd, F.D.; Skolnik, L.H.; Taylor, R.W.; Capone, B.R. & Roosild, S.A. (1978). Platinum Silicide Schottky-Barrier IR-CCD Image Sensors. *Int. Conf. on Application on CCD's*, San Diego, Oct. 25-27

Kosonocky, W.F.; Erhardt, H.G.; Meray, G.M.; Shallcross, F.V.; Elabd, H.A.; Cantella, M.; Klein, J.; Skolnik, L.H.; Capone, B.R.; Taylor, R.W.; Ewing, W.; Shepherd, F.D. & Roosild, S.A. (1980). Advances in Platinum-Silicide Schottky-Barrier IR-CCD Image Sensors. *SPIE Proc.*, Vol. 225, pp. 69-71

Kosonocky, W.F. ; Shallcross, F.V. & Villani, T.S. (1985). 160x244 Element PtSi Schottky-Barrier IR-CCD Image Sensor. *IEEE Trans. Electron Dev.*, vol. ED-32, No.8, pp. 1564

Kosonocky, W.F. (1991). Review of infrared image sensors with Schottky–barrier detectors. *Optoelectronics Devices and Technologies*, Vol.6, pp. 173–203

Kosonocky, W.F. (1992). State-of-the-art in Schottky-barrier IR image sensors. *SPIE Proc.*, Vol.1682, pp. 2–19

Kurianski, J.M.; Shanahan, S.T.; Theden, U.; Green, M.A. & Storey, J.W.V. (1989). Optimization of the cavity for silicide Schottky infrared detectors. *Solid-State Electronics*, Vol.32, pp. 97–101

Lee, M. K.; Chu, C. H.; Wang & Y. H. (2001) 1.55-μm and infrared-band photoresponsivity of a Schottky barrier porous silicon photodetector. *Opt. Lett.*, Vol.26, No.3, pp. 160-162

Liang, T.K.; Tsang, H.K.; Day, I.E.; Drake, J.; Knights, A.P. & Asghari, M. (2002). Silicon waveguide two-photon absorption detector at 1.5 μm wavelength for autocorrelation measurements. *Appl. Phys. Lett.*, Vol.81, No.7, pp. 1323

Liu, A.; Jones, R.; Cohen, O.; Hak, D. & Paniccia M. (2006). Optical amplification and lasing by stimulated Raman scattering in silicon waveguides. *J. Lightw. Technol.*, Vol.24, No.3, pp. 1440-1445

Liu, A.; Liao, L.; Rubin, D.; Nguyen, H.; Ciftcioglu, B.; Chetrit, Y.; Izhaky, N. & Paniccia, M. (2007). High-speed optical modulation based on carrier depletion in a silicon waveguide. *Opt. Express*, Vol.15, No. 2, pp. 660-668

Maier A. (2006). *Plasmonics: Fundamentals and Applications*. Springer, New York, USA

Michael, C.P.; Borselli, M.; Johnson, T.J.; Chrystal, C. & Painter, O. (2007). An optical fiber-taper probe for wafer-scale microphotonic device characterization. *Opt. Express*, Vol.15, No. 8, pp. 4745-4752

Muriel, M.A . & Carballar, A. (1997). Internal field distributions in fiber Bragg gratings. *IEEE Photonics technology letters*, Vol. 9, No.7, pp. 955, 1997

Olivieri, A.; Akbari A. & Berini, P. (2010) Surface plasmon waveguide Schottky detectors operating near breakdown. *Phys. Status Solidi RRL*. Vol 4, No.10, pp. 283 – 285

Reather, H. (1988). *Surface Plasmons on Smooth and Rough Surfaces and on Gratings*. Springer, Berlin, Germany

Rogalski, A. (1999). Assessment of HgCdTe photodiodes and quantum well infrared photoconductors for long wavelength focal plane arrays. *Infrared Phys. Technol.*, Vol.40, pp. 279–294

Rowe, L.K.; Elsey, M. ; Tarr, N.G. ; Knights, A.P. & Post, E. (2007). CMOS-compatible optical rib waveguides defined by local oxidation of silicon. *Electron. Lett.*, Vol.43, No.6, pp. 392-393

Scales, C. & Berini, P. (2004). Schottky Barrier Photodetectors, *U.S. Patent No. 7,026,701*.

Scales, C.; Breukelaar, I. & Berini, P. (2009). Surface-plasmon Schottky contact detector based on a symmetric metal stripe in silicon. *Opt. Lett.*, Vol. 35, No.4, pp. 529-531

Scales, C. & Berini, P. (2010). Thin-film Schottky barrier Photodetector Models. *IEEE Journal of Quantum Electronics*, Vol.46, No.5, pp. 633-643

Scales, C.; Breukelaar, I.; Charbonneau, R. & Berini, P. (2011). Infrared Performance of Symmetric Surface-Plasmon Waveguide Schottky Detectors in Si. *IEEE J. Lightw. Tech.*, Vol 29, No. 12, pp. 1852-1860

Sheperd, F.D.; Vickers, V.E. & Yang, A.C. (1971). Schottky Barrier Photodiode with a Degenerate Semiconductor Active Region. U.S. Patent No. 3.603.847

Shepherd, F.D. & Yang, A.C. (1973). Silicon Schottky Retinas for Infrared Imaging. *IEDM Tech. Dig.*, pp. 310-313

Shepherd, F.D.; Taylor, R.W.; Skolnik, L.H.; Capone, B.R.; Roosild, S.A.; Kosonocky, W.F. & Kohn, E.S. (1979). Schottky IRCCD Thermal Imaging. *Adv. Electron. Electron Phys.*, 7th *Symp. Photo-Electronic Image Devices*, Vol. 22, pp. 495-512

Shepherd, F.D. (1984). Schottky diode based infrared sensors. *SPIE Proc.*, Vol. 443, pp. 42–49

Shepherd, F.D. (1988). Silicide infrared staring sensors. *SPIE Proc.*, Vol.930, pp. 2–10

Shepherd, F.D. (1998). *Platinum silicide internal emission infrared imaging arrays.* Academic Press, New York, USA

Shiraishi, T.; Yagi, H.; Endo, K.; Kimata, M.; Ozeki, T.; Kama, K. & Seto, T. (1996). PtSi FPA with improved CSD operation. *SPIE Proc.*, Vol.2744, pp. 33–43

Sze, S.M. (1981). *Physics of Semiconductor Devices.* John Wiley & Sons, New York, USA

Taylor, R.W.; Skolnik, L.H.; Capone, B.R.; Ewing, W.; Shepherd, F.D.; Roosild, S.A.; Cochrum, B.; Cantella, M.; Klein, J.; Kosonocky, W.F. (1980). Improved Platinum Silicide IRCCD Focal Plane. *SPIE Proc.*, Vol. 217, pp. 103-110

Vickers, V.E. (1971). Model of Schottky Barrier Hot-electron-Mode Photodetection. *Applied Optics, Vol.* 10, No.9, pp. 219

Vivien, L.; Pascal, D.; Lardenois, S.; Marris-Morini, D.; Cassan, E.; Grillot, F.; Laval, S.; Fedeli, J.M. & El Melhaoui, L. (2006). Light injection in SOI microwaveguides using high-efficiency grating couplers. *J. Lightw. Technol.*, Vol.24, No.10, pp. 3810-3815

Xu, Q.; Manipatruni, S.; Schmidt, B.; Shakya, J. & Lipson, M. (2007). 12.5 Gbit/s carrier-injection-based silicon micro-ring silicon modulators. *Opt. Express,* Vol.15, No. 2,pp. 430-436

Yagi, H.; Yutani, N.; Nakanishi, J.; Kimata, M. & Nunoshita, M. (1994). A monolithic Schottky-barrier infrared image sensor with 71% fill factor. *Optical Engineering*, Vol.33, pp. 1454–1460

Yeh, P. (1988). *Optical Waves in Layerer Media.* Wiley Interscience Publication, New York, USA

Yuan, H.X. & Perera, G.U. (1995). Dark current analysis of Si homojunction interfacial work function internal photoemission far-infrared detectors. *Appl. Phys. Lett.*, Vol.66, No.17, pp. 2262-2264

Zhu, S.; Yu, M.B.; Lo, G.Q. & Kwong, D.L. (2008a). Near-infrared waveguide-based nickel silicide Schottky-barrier photodetector for optical communications. *Appl. Phys. Lett.*, Vol.92, No.8, pp. 081103

Zhu, S.; Lo, G.Q. & Kwong, L. (2008b). Low-cost and high-gain silicide Schottky-barrier collector phototransistor integrated on Si waveguide for infrared detection. *Appl. Phys. Lett.*, Vol. 93, No.7, pp. 071108

Zhu, S.; Lo, G.Q. & Kwong, D.L. (2008c) Low-Cost and High-Speed SOI Waveguide-Based Silicide Schottky-Barrier MSM Photodetectors for Broadband Optical Communications. *IEEE Phot. Tech. Lett.*, Vol. 20, No.16, pp. 1396-1398

3

UV-Vis Photodetector with Silicon Nanoparticles

J.A. Luna-López[1], M. Aceves-Mijares[2], J. Carrillo-López[1],
A. Morales-Sánchez[3], F. Flores-Gracia[1] and D.E. Vázquez Valerdi[1]
[1]Science Institute-Research Center for Semiconductor Devices-Autonomous
Benemérita University of Puebla
[2]Department of Electronics, National Institute of Astrophysics,
Optics and Electronics INAOE
[3]Centro de Investigación en Materiales Avanzados S. C.,
Unidad Monterrey-PIIT, Apodaca, Nuevo León,
México

1. Introduction

Nowadays, photodetector devices are important components for optoelectronic integration. In the past, various photodetector structures have been developed from pn junction, pin diode, bipolar transistor, avalanche photodiodes (APD), and metal-semiconductor-metal (MSM) structures (Ashkan et al., 2008; Hwang & Lin, 2005; DiMaria et al., 1984a, 1984b; Sabnis et al., 2005; Foster et al., 2006). In these structures, different semiconductors such as Si, III-V and II-VI compounds have been used, depending on the wavelength range to be detected. Nevertheless, silicon is the most common and important semiconductor in the integrated circuit technology, but it has an indirect band gap inhibiting optical functions. Silicon sensors are usually used in the visible-to near infrared (VIS-NIR) range. However, some commercial Si sensors have been enhanced to detect in the ultraviolet (UV) range, but they have one or more of the following weaknesses: very expensive, reduced responsivity in the VIS-NIR range, lack of compatibility with IC processes and a complex technology. So, most of the available materials for UV detection are not silicon but compound semiconductors (Hwang & Lin, 2005). Then, many works have been done to study Si-based optoelectronics materials to overcome the drawback of silicon. Silicon rich oxide (SRO) is one of such materials. SRO is a variation of silicon oxide, in which the content of silicon is changed. The main characteristic to enhance the UV detection in this material is the formation of silicon nanostructures. Among the different techniques to synthesize Si-nps in SRO films, the low pressure chemical vapour deposition (LPCVD) technique offers the films with the best luminescence properties, compared with those obtained by plasma enhanced CVD and silicon implantation into SiO_2 films (Morales et al., 2007).

The optical and electrical properties of SRO-LPCVD films have been extensively studied as a function of the silicon excess, temperature and time of thermal annealing (Morales et al., 2007, Zhenrui et al., 2008). Silicon nanoparticles (Si-nps) with different sizes in SRO-LPCVD films have been observed by transmission electron microscope (TEM) technique (Luna et al., 2009). Depending on the excess of Si content, SRO possesses some special properties such as charge trapping, carrier conduction and luminescence. Some novel devices have been proposed using these properties (Zhenrui et al., 2008, Luna et al., 2006, Berman et al., 2008). In Luna et al., 2006, a photodetector using SRO and an induced PN junction were presented, and it was shown that even when the active area is covered by opaque aluminium the photocurrent is considerable. However, it is not clear the reason of such a high response. Also, Aceves et al. proposed (Aceves et al., 1999) that an induced pn junction could be used as a photodetector and that if high resistivity silicon is used a high photoresponse could be expected. However, this idea has not been corroborated.

In this chapter, optical properties as absorption coefficient and optical band gap of SRO with Si-nps and different silicon excess were also studied in order to be able to apply them in optical photodetectors. A photodetector made of a simple Al/SRO/Si MOS-like grid structure is reported. The devices fabricated on high resistivity substrate possess high responsivity in photodetection from UV to VIS-NIR radiation. This structure requires few fabrication process steps and is completely compatible with the complementary metal-oxide-semiconductor (CMOS) technology. The possible mechanisms involved in the light detection and the extended photosensitivity of this structure are analysed and discussed.

2. Experimental procedure

The experimental procedure is divided in two parts. The first part is the following: SRO films were deposited on silicon substrates in a horizontal LPCVD hot-wall reactor using nitrous oxide (N_2O) and silane (SiH_4) as the reactant gases at 700°C. The partial pressure ratios Ro = $[N_2O]/[SiH_4]$, were Ro = 10, 20, and 30. The thicknesses of the SRO films were about 500 nm. Silicon Excess in these films are about 12.7, 8.0 and 5.5 at.% for Ro = 10, 20 and 30, respectively (Luna et al., 2007). After deposition, the films were densified at 1000 °C for 30 minutes in nitrogen atmosphere, and then some of them were thermally annealed at 1100°C for 180 minutes in nitrogen atmosphere.

The XRD measurements were done using a Bruker AXS D8 Discover diffractometer with Cu Kα radiation. The surface morphology of SRO films was studied using a nanosurf easy Scan AFM system version 2.3, operated in contact mode. A 4×4 μm^2 scanned area was used for each topographic image, and a 450-μm-long single-crystal Si cantilever operated at 12 kHz (type vista probes CL contact mode AFM probes) was used. Five different scans were done for each sample, showing good reproducibility. AFM images were analysed using scanning probe image processor (SPIP) software (Jogensen, 2002).

For TEM measurements, an energy-filtered TEM (EFTEM) was used. The images were obtained using an electronic microscope (JEOL JEM 2010F); all the EFTEM images were measured using a Si plasmon of 17 eV and cross-section views. EFTEM enables us to detect amorphous Si nanoclusters embedded in an oxide matrix (Spinella et al, 2005). Also, an FEI Tecnai F30 high-resolution TEM (HRTEM) with acceleration voltage of 300 kV and line resolution of 0.2 nm was used to study the microstructure of the Si nanocrystals.

Room-temperature photoluminescence (PL) of the SRO films was measured using a Perkin-Elmer spectrometer LS-50B model with a xenon source and a monochromator. The samples were excited using 250-nm radiation, and the emission signal was collected from 400 to 900 nm with a resolution of 2.5 nm. A cut-off filter above 430 nm was used to block the light scattered from the source.

Optical transmittance measurements were made in the UV range to near infrared at room temperature, using a spectrophotometer Perkin-Elmer LMBD 3B UV/VIS.

In the second part of the experimental procedure SRO films were deposited on n-type Silicon (100) substrates with resistivity of 2000-5000 Ω-cm and with a N+ implanted region on the back side. SRO layers were deposited in a horizontal LPCVD hot wall reactor using SiH$_4$ (silane) and N$_2$O (nitrous oxide) as reactive gases at 700 °C. The gas flow ratio Ro was used to control the amount of silicon excess in the SRO films. Ro = 10, 20, and 30, corresponding to a silicon excess from 12 to 5 %, were used for this experiment. After deposition, the samples were thermally annealed at 1000° C in N$_2$ atmosphere for 30 minutes. Aluminium grids with area of A = 0.073 cm^2 were patterned on the SRO surface by evaporation and standard photo-lithography. Figures 1(a) and 1(b) show a cross-section view scheme, with an approximation at only 3 grid electrodes and an image of the surface of the fabricated devices. The surface Al grid electrodes have several fingers with 40 μm width and distanced at 205 μm, as observed in figure 1(b). It is important to remark that the zones between fingers are covered by the SRO film.

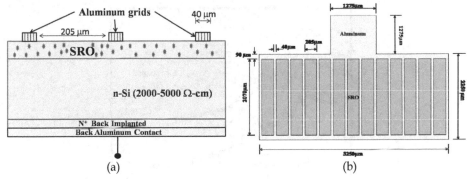

(a)   (b)

Fig. 1. Scheme of the Al/SRO/Si MOS-like device, (a) a cross-section view scheme of the fabricated devices with an approximation at only 3 grid electrodes and (b) dimensions of a device and of the grids electrodes.

Ellipsometric measurements were made with a Gaertner L117 ellipsometer to obtain the thickness and refractive index of the SRO films before annealing, whose values are shown in Table 1.

Ro	Refractive index	Thickness (Å)
10	1.78 ± 0.01	720 ± 28
20	1.55 ± 0.03	755 ± 25
30	1.46 ± 0.01	591 ± 3

Table 1. Refractive index and thickness of SRO films.

Current versus voltage (I-V) measurements were performed at room temperature, illuminated with UV or white light and under dark conditions, using a computer controlled Keithley 6517A electrometer. The voltage sweep was done with a rate of 0.1 V/s. Illumination was performed with an UV lamp (UVG-54, 5 to 6 eV approximated range) and a white light lamp (1.7 to 4 eV approximated range) with an output power of 4.3 mW/cm^2 and 3.2 mW/cm^2, respectively. The power of lamps was measured using a radiometer (International Light, USA. IL1 400A). On the other hand, other current versus voltage (I-V) measurements were performed at room temperature using a computer controlled Keithley 236 source-meter, under illuminations at specific wavelengths of 400, 650 and 1000 nm conditions, with a Sciencetech 9040 monochromator. The measured power were 83.6 $\mu W/cm^2$, 238.9 $\mu W/cm^2$ and 11.5 $\mu W/cm^2$ for 400, 650, 1000 nm wavelengths, respectively.

3. Results

3.1 Structural, optics and photoelectric results

Figure 2 shows the XRD patterns of SRO films with Ro = 10 and 20, annealed at 1100 °C for 180 minutes on silicon substrate. Peaks at 21, 28.5° and 47.4° are observed for SRO films with Ro = 10 and they are ascribed to SiO_2, Si (111) and (220), respectively. However, those peaks at 28.5° and 47.4° disappeared for Ro = 20. As Ro becomes higher (Ro = 30), the diffraction peaks were not observed (not shown here).

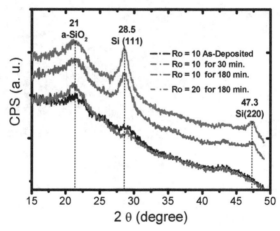

Fig. 2. XRD patterns of SRO films with Ro = 10 and 20 annealed at 1100 °C for 180 minutes on silicon substrate.

The average crystal size (D), dislocation density (δ) and microtension (ε) of the silicon nanocrystals (Si-ncs) for Ro = 10 were estimated from the Si (111) peak using the Scherrer formula (Karunagaran, B., et al., 2002, Comedi,D., et al., 2006,) and are listed in Table 2. δ and ε decreased when the annealing time increases for SRO_{10} films, which is related to a reduction on the defect concentration and imperfections of the nanocrystal cell (Yu et al, 2006).

		Samples of SRO on silicon			
SRO films	2θ (°)	[hkl]	(D) (nm)	δ (10^{12} cm^{-2})	ε (10^{-3})
10TT30	28.65	Si [111]	4.6	4.6	8.0
10TT180	28.55	Si [111]	5.4	3.4	7.0
20TT180	29.1	Si [111]	2.8	12.2	13.0

Table 2. Structural parameters of the SRO Films obtained from the XRD measurements for Si (111) peak.

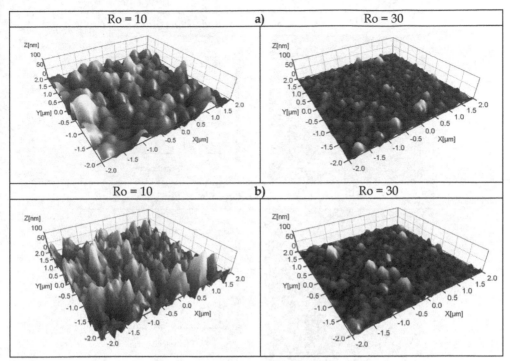

Fig. 3. 3D AFM images of a) as-deposited and b) thermally annealed at 1100 °C SRO_{10} and SRO_{30} films on silicon. Scanned area: $4 \times 4 \ \mu m^2$.

Figure 4 shows a comparison of the average roughness <Sa> for the SRO films with different Si excess for as-deposited, densified and annealed films.

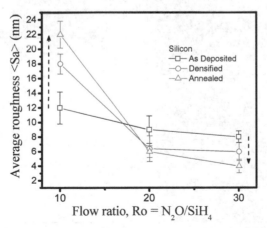

Fig. 4. Average roughness <Sa> as a function of the flow ratio (Ro) for as-deposited, densified, and thermally annealed SRO films deposited on silicon (lines are plotted as an eye-guide). Scanned area: $4 \times 4 \ \mu m^2$.

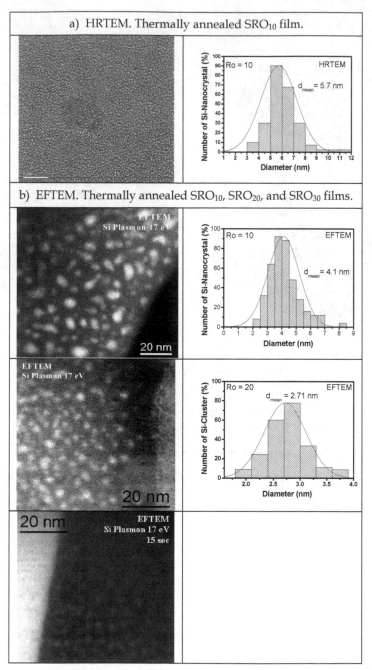

Fig. 5. Silicon nanocrystals in SRO_{10} films and silicon clusters in SRO_{10}, SRO_{20}, and SRO_{30} films were observed by a) HRTEM and b) EFTEM, respectively, after thermal annealing at 1100 °C.

Figure 3 displays three-dimensional (3D) AFM images of as-deposited and thermally annealed SRO films with different excess silicon. All images exhibit a rough surface. For as-deposited SRO films, the roughness decreases with increasing Ro, as shown in Fig. 3 (a). After thermal annealing, roughness increased for SRO films with Ro = 10, as depicted in Fig. 3 (b); for Ro = 20 and 30, no significant influence is visible from 3D images.

Figure 5 shows the HRTEM and EFTEM images of the annealed SRO films. For Ro=10, HRTEM and EFTEM images clearly show Si nanocrystals (Si-ncs) with a mean size of ~ 5.7 nm, as seen in Figures 5 (a) and 5 (b). Si-ncs were not observed for Ro = 20 and 30. However, bright zones associated with amorphous silicon clusters were observed by EFTEM on SRO films with Ro = 20 and 30, as shown in Fig. 5 (b). Si clusters with a mean size of ~ 2.7 nm were observed for Ro = 20, whereas for Ro = 30, because of low contrast and sharpness, the shape was not very clear, but Si clusters with size ≤ 1 nm could be identified.

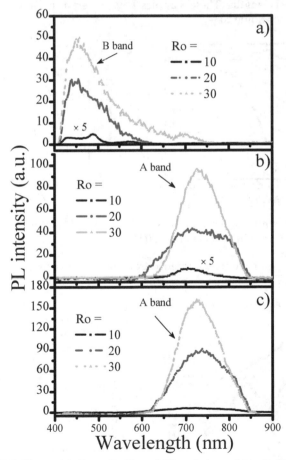

Fig. 6. PL spectra of SRO films with Ro = 10, 20 and 30 and a) as- deposited, b) densified at 1000 °C for 30 min, and c) thermally annealed at 1100° C for 180 minutes deposited on silicon substrate.

Figures 6 (a)–6 (c) show the PL spectra of the as-deposited, densified, and annealed SRO films, respectively. Two PL bands can be observed: B band (400–600 nm) and A band (600–850 nm). For as-deposited SRO films, the intensity of the B band increases with Ro, as shown in Fig. 6 (a). After densification, the B band disappears and the A band becomes visible [see Fig. 6 (b)]; prolonged annealing at higher temperature of 1100 °C for 180 min enhanced its intensity, as shown in Fig. 6 (c). SRO films with large Ro (low excess Si) always show higher PL intensity. On the other hand, the PL peak position did not show any significant shift with Ro.

Transmittance spectra were measured to obtain the optical band gap of SRO films. Transmittance spectra for Ro = 10, 20 and 30 films deposited on sapphire are shown in Figure 7. As can be observed, the transmittance of all these films is relatively high (> 80 %) between 600 and 900 nm, and reduces to zero for wavelengths below 600 nm. The annealing time does not produce a clear variation; however, the Ro causes a clear shift of the curves towards lower wavelengths. These samples have different behaviours, in agreement with the Ro value and annealing time.

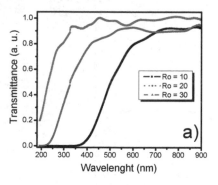 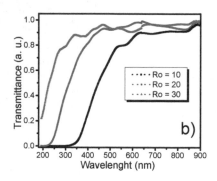

Fig. 7. UV/Vis transmittance spectra for Ro = 10, 20, and 30 on sapphires substrates. a) densified at 1000 °C and b) annealed at 1100 °C.

The absorption coefficients were determined from transmission spectra of Figure 7, and they are shown in Figure 8. The absorption coefficients show different behaviour with the Ro values.

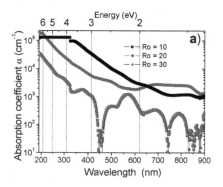 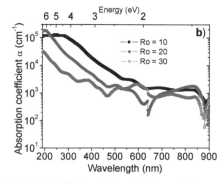

Fig. 8. Absorption coefficients in the range of the U-V to NIR for a) densified and b) thermally annealed SRO films.

The fundamental absorption edge in most semiconductors follows an exponential law (Pankove, 1975). The absorption coefficient α is correlated with the transmittance T and the reflectance R of a sample with thickness d through the relation:

$$T \approx (1-R)^2 e^{-\alpha d} \tag{1}$$

The absorption coefficient α shown in Figure 8 was estimated as:

$$\alpha(\hbar v) = \frac{-\ln(T(\hbar v))}{d} \tag{2}$$

From the Tauc law:

$$(\alpha \hbar v)^{1/2} \propto (\hbar v - E_g) \tag{3}$$

Graphs of $(\alpha \hbar v)^{1/2}$ versus photon energy ($\hbar v$) and the corresponding optical band gap are shown in Figure 9.

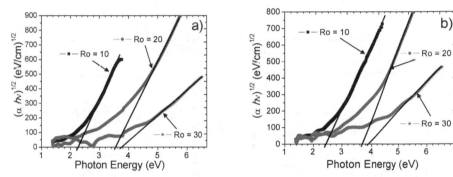

Fig. 9. $(\alpha \hbar v)^{1/2}$ vs. photon energy a) densified and b) thermally annealed SRO films with different Ro. Straight lines are used to estimate the optical band gap.

It is observed that the position of the absorption edge moves towards higher energy when Ro increases. The values obtained for the optical band gap are listed in Table 3.

Ro	Densified (1000°C)	Band gap (Eg)		
		Annealed at 1100°C		
		30 min	60 min	180 min
10	2.28±0.04	2.4±0.04	2.4±0.02	2.43±0.04
20	3.52±0.03	3.57±0.03	3.6 ±0.06	3.69 ±0.03
30	3.76 ±0.07	3.73 ±0.06	3.86±0.04	3.89±0.04

Table 3. Energy of the optical bang gap obtained from the Tauc's method for SRO films with different Ro values and times of thermal treatment.

The optical band gap increases with Ro. In this case, as the silicon excess reduces the absorption coefficient also reduces, as shown in Figure 8. As Ro moves towards low values, the band gap moves towards that of the bulk silicon; consequently, it is expected that the properties of an indirect semiconductor be kept.

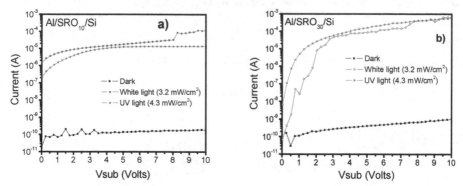

Fig. 10. I-V curves of the Al/SRO/Si MOS-like grid structure for a) SRO_{10} and b) SRO_{30} under dark and illumination conditions.

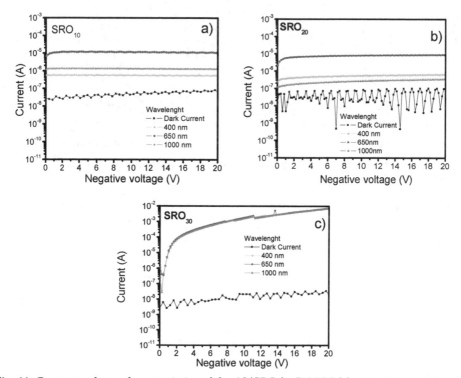

Fig. 11. Current-voltage characteristics of the Al/SRO/n-Si MOS-like structures with a) SRO_{10}, b) SRO_{20}, and c) SRO_{30} films under dark and illuminated with 250, 400, 650, 1000 nm wavelengths.

MOS-like structures were fabricated and measured. Figures 10(a) and 10(b) show, respectively, the typical I-V characteristics of devices with Ro = 10 and 30 at the surface inversion condition (negative voltage at the gate respect to the substrate), in dark and under UV and white light illumination. The dark current is on the order of 10^{-10} A, indicating a low leakage current in these structures. However, a large photocurrent was obtained when the structures were illuminated, this fact indicative of a high optical sensitivity in this simple structure. Similar photo-response was obtained with structures made with SRO of Ro = 20.

I-V characteristics of devices with Ro = 10, 20 and 30 at the surface inversion condition and under illumination at specific wavelengths are shown in Figures 11(a), 11(b) and 11(c), respectively. The order of the dark current is 10^{-8} A indicating a low leakage current in these structures. This current value changed from that obtained in Figure 10 because of the fact that different devices were measured and these ones have a bigger dark current. Nevertheless, at both cases a large photocurrent was obtained when the structures were illuminated. Therefore, these simple structures show high optical sensitivity. Different photocurrents were obtained for the different Ro's in each wavelength, being the SRO structures with Ro = 30 which have the highest photo-response.

4. Analysis and discussion

The results show that SRO films properties depend on the silicon excess and the annealing conditions. XRD results showed that in SRO films with thermal annealing, diffusion of the silicon excess took place creating silicon nanoparticles indicated mainly by the peaks, and their size depends upon the silicon excess in the SRO films. Diffractograms on Figure 2 show crystalline orientations (111), (220) and (311) of the nc-Si. It is clear that, when the silicon excess is higher, the agglomerated Si is present and with thermal annealing the generation of nc-Si is obtained, the diffusion coefficient of Si in the SRO at 1100 °C was calculated as $\sim 1 \times 10^{-16}$ cm^{2}s^{-1}. Diffraction peaks indicate the formation of nc-Si and it should be interpreted as a phase separation between nc-Si and SiO$_2$ induced by thermal annealing.

Si-ncs sizes obtained by XRD are similar to those observed by HRTEM and EFTEM. For Ro = 10, HRTEM images clearly show Si-ncs with a mean size of 5.7 nm but Si-ncs were not observed for Ro = 20 and 30. However, bright zones associated with amorphous silicon clusters were observed by EFTEM on SRO films with Ro = 20 and 30. Si clusters with a mean size of 2.7 nm were observed for Ro = 20, whereas for Ro = 30 the shape was not very clear, because of low contrast and sharpness, but Si clusters with 1 nm size could be identified.

AFM images of the SRO films on silicon substrates show that the shape and size of the roughness change with Ro. For SRO films with Ro = 10, the roughness increased after thermal annealing. However, for SRO with Ro = 20 and 30, the roughness reduced, as shown in Figure 4. For Ro = 10 the roughness can be as high as a maximum of 24 nm. However, for Ro = 30 the surface looks smooth and the maximum height is about 5 nm. Then, for Ro = 10, the roughness observed through AFM can be associated to Si-ncs. Nevertheless, for SRO films with Ro = 20 and 30 the roughness observed can be ascribed to silicon compounds, or silicon clusters, rather than to Si-ncs. Therefore, elemental Si, SiOx, and SiO_2 phases separate after annealing, and depending on the silicon excess, one of those phases can be dominant.

Si clusters in the SRO films are produced by diffusion of the Si excess at high temperature. That is, when SRO is annealed, the silicon particles diffuse creating silicon agglomerates around a nucleation site. If the Si excess is high enough, the Si clusters will be crystallized forming Si-ncs. However, when the Si excess reduces, Si clusters are amorphous due to the large lattice mismatch between SiO_2 and Si. If the Si excess is low enough, there are no pure Si clusters created, instead, Si-O compounds are formed in SRO films. Then a more homogeneous material is obtained. Of course, these two mechanisms are not exclusive each other, and they can exist simultaneously. However, one of them will dominate depending on the silicon excess.

On the other hand, there are different factors that have an important impact on the optical properties of the SRO films. Therefore, it is necessary to understand these factors for designing materials to be used in photodetectors. In our case, factors such as substrate type, silicon excess, annealing temperature and composition are very important for the optical properties of SRO films. PL also depends on Ro and thermal-annealing conditions. In Figure 6, all the as-deposited SRO films show a weak PL band at 400–600 nm (B band). After thermal annealing, the B band disappears and another PL band (A band) dominates. The intensity of the A band increases with the time and annealing temperature. The PL property of SRO has been extensively studied in literature. Two major mechanisms for PL in this kind of materials are generally accepted: quantum confinement effects in the Si-ncs and defect-related effects. In our samples, intense PL (A band) in SRO_{30} was observed; however, SRO_{30} has silicon compounds rather than Si-ncs. Therefore, the A band cannot be associated with quantum confinement effects; instead, it should be associated with Si-related defects. As reported previously (Lee et al., 1979, Lin et al., 2005), the B band is also associated with F and E defects and different types of oxygen vacancies. Although there are many reports about the origin of PL, most of them relates PL with the bulk microstructures of SRO. A very few studies report on the correlation between the PL and the surface morphology of the SRO films. Other authors (Torchynska et al, 2002) have reported that the diameter and area of the grains on the porous silicon film surface have an important influence on the PL spectra. In our case, SRO is not porous Si but it is considered to be Si nanoclusters embedded in an oxide matrix with very stable optical and electrical properties—quite different than porous silicon. However, from AFM images, the surface is completely irregular, with bumps of different heights and shapes. So, these bumps can be treated as grains on the surface. The grains vary in size depending on the Ro value and annealing time. They can then be related to the emission, as it was done for porous silicon. The height of the roughness, equivalently to the height of the grains, was already related to the silicon cluster size (Luna et al, 2007).

It is observed that the main influence on the structural and optical properties is due to the Ro value of the SRO films, their thickness, annealing time and substrate type. In XRD and AFM measurements important changes in the structure were observed, which indicates that the substrate has influence on the structural properties of the deposited SRO films. In the PL spectra the unusual high intensity of the A band is remarkable. The explanation for this observation is that different oxidation mechanisms lead to different termination of the tetrahedron SiO_4 and Si_4 of the silicon oxide and silicon, respectively. Si can be terminated with different combinations of $Si-Si_{4-n}O_n$, compounds (with n = 0 to 4), it can be terminated by double bond to an oxygen atom (Si=O), oxygen vacancies or by a bond to a hydroxyl group (Si-O-H).

Taking these results into account, it seems that the types of radiative centres producing the luminescence in SRO films deposited on silicon substrate are similar in the A band, but different in the B band. Therefore, depending on the substrate type, specific structural defects are formed in the SRO films. The Si content in the SRO system to form the phase (crystalline or amorphous) and the structure (crystal size and morphology) of the Si-nps, and especially of their surrounding $Si-SiO_2$ interface, has significant influence on the optical properties. Also, the optical band gap tends towards the optical band gap of the Bulk silicon by increasing the silicon excess, and with a lower silicon excess the optical band gap will tend towards that of the SiO_2.

The large variety of factors that can be used to influence the structural and optical properties makes these materials very interesting for optoelectronics applications. A shift of the optical band gap is possible by varying the silicon excess. Also, the PL emission can be tuned by thermal annealing, silicon excess and substrate type.

Therefore, MOS-like structures were designed to obtain high photocurrent. The results show that Al/SRO/Si grid structures have high photocurrent at different wavelengths. The large photocurrent is due to the photocarriers generation in the SRO films and also in the Si substrate. The mechanism of the carrier transport in the SRO layers can be via tunneling (Yu et al, 2008). High density of Si-nps and defects in the SRO films create conduction paths and then the photogenerated carriers can move through them allowing a large photocurrent.

This simple structure presents high optical sensitivity for UV and the visible to near infrared range (from 250 nm to 1000 nm), as depicted in Figures 10 and 11. The UV response of the SRO layer has been demonstrated in different reports, apparently the silicon nanoparticles (embedded or not) are in some way sensible to the UV. In (Nayfeh et al, 2004) the authors use silicon nanocrystals (not embedded) to develop an UV sensor in the 250 nm to 350 nm range. Also, our group has developed an UV-Vis-NIR sensor using SRO (with embedded Si nps) (Berman et al., 2008). Moreover, it is well known that red emission is obtained when SRO is irradiated with UV. So, UV response of the SRO could cooperate in two ways to the observed response of the devices reported in this work. First, the red photoluminescence produced by the SRO impinges into the silicon, which has a high sensitivity to these wavelengths. Second, the SRO itself has demonstrated some conduction properties under illumination, so this current component could be added to the photocurrent (Luna et al, 2010).

Now, the high photoresponse can only be explained if the whole device is considered photosensible. That includes the components due to the SRO (already discussed), and the components in the silicon. As Figure 12 shows, there are two zones in the silicon substrate where the light could produce electron-hole pairs (EHP): (1) under the Al fingers and (2) in the exposed regions between the fingers. Due to the bias of the device, under the Al fingers the silicon surface is in inversion (layers of holes formed), forming an induced-PN junction. However, owing to the fact that this PN junction is covered by an opaque Al layer, very few photons will produce photocurrent; then the current component due to this layer should be small. So, the PN junctions could act as a collector of the minority carrier produced somewhere else, and then contribute to the photocurrent through the SRO that allows the electrons to move from the substrate to the Al electrode.

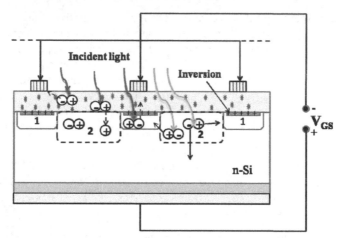

Fig. 12. A cross-section scheme of the zones in the SRO films and silicon high resistivity substrate where the light could produce electron hole-pairs (EHP). (1) Under the Al fingers the silicon surface is in inversion due to the bias of the device, in this case only a few photons arrive to the edge of the region, and (2) in the SRO films in the Al uncover between the fingers. The dash lines marks the maximum length where a photon can reach.

On the other hand, the SRO traps charge both positive and negative, so that the surface in this zone could be in accumulation if the SRO is positively charged, in inversion if the SRO is negatively charged, or neutral if there is no charge in the SRO. The resistivity of the silicon substrates used for this experiment is 2000-5000 ohms-cm, so small quantities of trapped charge could bias the surface. Moreover, the charge could get to be trapped during the fabrication process, especially during the high temperature process.

The inversion or accumulation of the surface produces channels where the carriers could move toward the fingered zone where they contribute to the photocurrent. However, in our devices the dark current is very low (< 10^{-8} A) meaning that the channels do not exist.

Then, the only possibility to increase so much the photocurrent is that the uncovered neutral region under illumination produces electron-hole pairs able to arrive at the Al covered area. As known, only the EHP produced within a diffusion length, $L = \sqrt{D\tau}$ (D is the diffusivity and τ is the lifetime), will arrive at the PN junctions. In this case, only the holes which are the minority carriers are significant. The diffusivity can be estimated from the mobility using the Einstein's relationship. Also, the resistivity of the silicon substrate used here is very high and the mobility (510 cm²/Vs) does not vary much for resistivities greater than 100 Ω-cm (Cronemeyer, 1957). Therefore, a conservative value of the diffusivity is 13.184 cm²/s (Grove, 1967). Härkönen et al. (Härkönen et al., 2006) studied the lifetime of high resistivity silicon under low and high injection levels of carriers. They found that for high resistivity silicon the recombination lifetime is longer than 1 ms. Just for reference in our group, generation lifetime of high resistivity silicon was estimated to be between 1 and 2 ms (Luna et al., 2005). Then, a value of diffusion of 0.114 cm, or 1140 μm is calculated.

Now, for UV radiation (200 - 300 nm) the absorption coefficient is approximately, 1×10^6 cm^{-1}, and the absorption depth is 0.01 μm (Sze, 2006). For other wavelengths, for example around 600 nm, the absorption depth is about 5 μm. In spite of fact that the absorption is quite superficial for UV, most of the generated holes reach the induced PN junctions due to their long diffusion length, and in general this is true for all the detection range: UV-Vis-NIR. So, the high sensitivity observed in this device is due to a large silicon volume where all the generated carriers contribute to the photocurrent and, as a complement, some photo-effects in the SRO also contribute to the photocurrent.

The structure with Ro = 30 produces the highest current. However, it is well known that the lower Ro values increase the SRO conductivity, so if the current goes through the SRO, the more conductive structure will be that with Ro = 10. On the other hand, it is also known that silicon excess of 5% (for Ro = 30) produces the highest PL emission (Morales et al., 2005, Luna et al., 2009), and then the more sensible to radiation, where the PL (red emission) generates e-h pares in the induced PN junction. Then, these two mechanisms have to be further studied to determine which one is more adequate to improve the photoresponse of our sensor.

5. Conclusion

The influence of different factors on the structural (Si-nc, a-Si, Si-nanocluster, roughness) and optical properties (PL, absorption coefficients, optical band gap) of the SRO films has been discussed from experimental data. PL increased with the annealing time and decreased with a greater amount of silicon excess. In our case, it has been discussed that PL is not due to Si-ncs. Si-clusters are the main nanoparticles that produce maximum PL due to the different type of defects. The formation and growth of Si-ncs and the formation of a-Si in samples with different silicon excess after different annealing times were investigated with XRD and HRTEM. Interestingly, it was found that the optical band gap strongly correlates with the silicon excess and it is shifted over a large energy range via the reactants ratio.

These interesting properties of the SRO films were used to fabricate MOS-like structures, and an Al/SRO/Si MOS-like grid structure was fabricated and tested as photodetector. Silicon-rich oxide films with different silicon excesses were used as dielectric layers and deposited on silicon substrates of very high resistivity. This structure shows high optical sensitivity for the whole UV-Vis-NIR range. The high photocurrent results from three different components. One of them is due to photo-carriers generated in the SRO films; another one comes from the red light that is photo generated in the SRO, as mentioned above. This red light impinges into the high resistivity silicon producing electron-hole pairs. The third component comes from the electron-hole pairs generated in the silicon by the light absorbed in the silicon volume. In spite of the fact that the UV radiation produces electron-hole pairs near the silicon surface, they contribute to the photocurrent because there is not a death zone. Due to the high resistivity of the silicon a long diffusion length is obtained, and the photo generated minority carriers are able to reach the induced PN junctions under the Al grids.

6. Acknowledgment

This work has been partially supported by CONACyT-154725, PROMEP-231 and VIEP-BUAP-2011. The authors acknowledge technicians of INAOE laboratory for their help in the samples preparation and measurements.

7. References

Aceves, M., O. Malik, V. Grimalsky, Application of silicon Rich Oxide Films in New Optoelectronic Devices, Physics and Chemistry of Solid State 2004, 5(2), pp. 234-240.

Aceves, M., W. Calleja , C. Falcony, J. A. Reynoso–Hernández, The Al/Silicon Rich Oxide/Si P-N Induced Junction As A Photodetector, Revista Química Analítica, 1999, 18, 102.

Ashkan, B.; Jason, J.; Yongho, C.; Leila, N.; Gunham, M.; Wu, Z.; Andrew, G.; Kapur, P. ; Kriehna, C. & Ant, U. (2008). Metal-Semiconductor-metal photodetectrors based on single-walled carbon nanotube film-GaAs Schottky contacts, J. Appl. Phys., Vol.13, (June 2008), pp. 114315, ISSN 0021-8979

Berman D. M., M. Aceves-Mijares, L. R. Berriel-Valdos, J. Pedraza, A. Vera-Marquina, Fabrication, characterization, and optimization of an ultraviolet silicon sensor, Optical Engineering, 2008, 47(10), 104001.

Chen Y., T. Cheng, . Cheng, C. Wang, C. Chen, C. Wie, Y. Chen, Highly sensitive MOS photodetector with wide band responsivity assisted by nanoporous anodic aluminum oxide membrane, Optics Express 2010, 18(1), pp. 56-62.

Comedi, D., O.H.Y. Zalloum, E. A. Irving, J. Wojcik, T. Roschuk, M. J. Flynn, and P. Mascher, X-Ray-Diffraction study of crystalline Si nanocluster formation in annealed silicon-rich silicon oxides, J. Appl. Phys. 99, 023512 (2006).

Comedi, D., Zalloum, O.H.Y., Irving, E. A., Wojcik, J., Roschuk, T., Flynn, M. J., and Mascher, P., (2006), X-Ray-Diffraction study of crystalline Si nanocluster formation in annealed silicon-rich silicon oxides, J. Appl. Phys. 99, 023512 (2006), ISSN 0021-8979

Di Maria, D. J., Kirtley, J. R., Pakulis J., Dong, D. W., Kuan, D., Pesavento, F. L., Theis, T. N., Cutro, N., Brorson, S. D.; (1984). Electroluminescence studies in silicon dioxide films containing tiny silicon islands, J. Appl. Phys.,Vol. 56(2) (1984), pp. 401-416, ISSN 0021-8979.

DiMaria, D. J., Dong, , Pesavento, D. W., F. L.; Enhanced conduction and minimized charge trapping in electrically alterable read-only memories using off-stoichiometric silicon dioxide films, J. Appl. Phys., 1984, 55(8), pp. 3000-3019, ISSN 0021-8979.

Donald C., Cronemeyer, Hall and Drift Mobility in High-Resistivity Single-Crystal Silicon, Phys. Rev., 1957, 105(2), 522.

Foster J., J. K. Doylend, P. Mascher, A. P. Knights, and P. G. Coleman, Optical attenuation in defect-engineered silicon rib waveguides, J. Appl. Phys., 2006, 99, pp. 073101, ISSN 0021-8979.

Grove, A. S., Physic and technology of semiconductor devices, John Wiley and sons, USA 1967.

Harkonen, J., E. Tuovinen, Z. Li, P. Luukka, E. Verbitskaya, V. Eremin, Recombination lifetime characterization and mapping of silicon wafers and detectors using the microwave photoconductivity decay (mPCD) technique, Materials Science in Semiconductor Processing, 2006, 9, 261.

Hwang, J. & Lin, C. (2005). Gallium nitride photoconductive ultraviolet sensor with a sputtered transparent indium–tin–oxide ohmic contact, Thin Solid Films, Vol.491 (May 2010), pp. 276-279, ISBN 842-6508-233.

Jøgensen, J. F., The Scanning Probe Image Processor (SPIP), Denmark, 2002, www.imagemet.com.

Karunagaran, B., Rajendra, R. T., Mangalaraj, D., Narayandass, K., Mohan, G., (2002), Influence of thermal annealing on the composition and structural parameters of DC magnetron sputtered titanium dioxide thin films, Cryst. Res. Technol., Vol. 37 (12), 2002, pp. 1285-1292.

Luna López, J. A., A. Morales Sanchez, M. Aceves-Mijares, Z. Yu, C. Dominguez, Analysis of surface roughness and its relationship with photoluminescence properties of silicon-rich oxide films, J. Vac. Sci. Technol. A, 2009, 27(1), 57.

Luna-López, A., M. Aceves-Mijares, O. Malik, R. Glaenzer, Low-and high-resistivity silicon substrate characterization using the Al/silicon-rich oxide/Si structure with comparison to the metal oxide semiconductor technique, J. Vac. Sci. Technol. A, 2005, 23 (3), 534-538.

Luna-López, J. A., Aceves M., O. Malik, Z. Yu, A. Morales, C. Dominguez and J. Rickards, Revista Mexicana de Física, S 53(7), 293 (2007).

Luna-López, J. A., M. Aceves-Mijares and O. Malik, Optical and electrical properties of silicon rich oxide films for optical sensor, Sensors and actuators A, 2006, 132, 278.

Luna-López, J. A., M. Aceves-Mijares, A. Morales-Sánchez, J. Carrillo-López, Photoconduction in silicon rich oxide films obtained by LPCVD, J. Vac. Sci. Technol. A 28(2), Mar/Apr 2010.

Morales A., J. Barreto, C. Domínguez, M. Riera, M. Aceves and J. Carrillo, (2005), Comparative study between silicon-rich oxide films obtained by LPCVD and PECVD, Physica E: Low-dimensional Systems and Nanostructures, 2007, 38(1), 54.

Nayfeh, O. M., S. Rao, A. Smith, J. Therrien, and M. H. Nayfeh, Thin Film Silicon Nanoparticle UV Photodetector, IEEE Photonics Technology Letters, 2004, 16(8), 1927.

Pankove, J. I., Optical Processes in Semiconductors, (Dover New York, 1975), Chap. 4-6.

Sabnis, V. A., Demir, H. V., Fidaner, O, Zheng, J. F., J S Harris, D A B Miller, N Li, T C Wu, H T Chen, Y M Houng (2005), Intimate Monolithic Integration of Chip-Scale Photonics Circuits, IEEE J. selected Topics in Quantum Electronics 2005, 11(6), pp. 1255.

Spinella C., C. Bongiorno, G. Nicotra, and E. Rimini, Appl. Phys. Lett. 87, 044102 2005.

Sze, S. M., Physics of semiconductor devices, John Wiley and sons, USA, (2006).

Torchynska, T. V., J. Appl. Phys. 92, 4019 2002.

Yu Z., M. Aceves-Mijares, J. A. Luna Lopez, J. Deng, Nanocrystalline Si-based metal-oxide-semiconductor, Proc. SPIE 7381 2009, pp. 7381 1H.

Yu Z., Mariano Aceves-Mijares, A. Luna-López, Enrique Quiroga and R. López Estopier, "Photoluminescence and Single Electron Effect Of Nanosized Silicon Materials". In:

Focus on Nanomaterials Research, Editor: B. M. Carota, ISBN 1-59454-897-8 Nova Science Publishers, Inc., 2008, 233.

Yu, Z., M. Aceves-Mijares, K. Monfil, R. Kiebach, R. Lopez-Estopier, J. Carrillo, Room temperature current oscillations in naturally grown silicon nanocrystallites embedded in oxide films, J. Appl. Phys., 2008, 103, 063706.

4

High Sensitivity Uncooled InAsSb Photoconductors with Long Wavelength

Yu Zhu Gao

College of Electronics and Information Engineering, Tongji University, Shanghai
China

1. Introduction

It is very attractive to research infrared (IR) materials and detectors in the 8-12 μm wavelength range for photodetection applications. Currently, HgCdTe is the dominant material system in this wavelength range. However, it suffers from chemical instability and nonuniformity due to the high Hg vapor pressure during its growth (Kim et al., 1996). Among III-V compound semiconductors, InAsSb alloy with a band gap as small as 0.1 eV has the advantages of high electron and hole mobilities, good operating characteristics at high temperatures, and high chemical stability. Therefore, the InAsSb system is a very promising alternative long-wavelength IR material to HgCdTe.

However, the lattice mismatch between InAsSb epilayers and the binary compound substrates is rather large (for InAs is larger than 6%, for GaAs is 7.2 ~ 14.6%). Thus it is very difficult to grow high-quality InAsSb single crystals with cutoff wavelengths of 8-12 μm using conventional technologies (Kumar et al., 2006). Narrow-gap InAsSb epilayers have been grown by molecular beam epitaxy (MBE) (Chyi et al., 1988), metalorganic chemical vapor deposition (MOCVD) (Kim et al., 1996), and liquid phase epitaxy (LPE) (Dixit et al., 2004). The thicknesses of these epilayers are about 2-10 μm. The dislocation densities observed in these thin films are as high as the order of 10^7 cm^{-2} that are caused by a large lattice mismatch (Kumar et al., 2006). It markedly lowers the terminal performance of the detectors.

A melt epitaxy (ME) method for growth of InAsSb single crystals on InAs and GaAs substrates with the wavelengths longer than 8 μm was proposed for a first time by Gao et al (Gao et al., 2002, 2006). The thickness of InAsSb epilayers reaches several decades ~ 100 μm. This thickness effectively eliminates the effect of lattice mismatch and results in a low dislocation density (the order of 10^4 cm^{-2}) in epilayers with a lattice mismatch larger than 6%. Based on the thick InAsSb epilayers grown by ME, high-sensitivity uncooled photoconductors with long wavelength were successfully fabricated (Gao et al., 2011). The IR photodetectors operating at room temperature need not coolers, thus have the important advantages of high speed, small volume, and good reliability. The response speed of them is more than three orders of magnitude faster than that of thermal detectors.

2. Melt epitaxy

We prepared InAsSb epilayers in a standard horizontal LPE growth system with a sliding fused silica boat in high-purity hydrogen ambient. Fig. 1 (a) shows the slideboat schematic

(Gao et al., 2002). The bottom of the tail part of the melt holder for the fused silica boat is flat. The original melt composition was $InAs_{0.04}Sb_{0.96}$. Seven Newtons of Sb, In, and non-doped InAs polycrystalline were employed as the source materials for the melt. The substrates were (100) oriented n-InAs and semi-insulating GaAs substrates. The growth melt of ME is not diluted, thus the composition of the solid is less different from that of the melt. We measured the composition of InAsSb solids by electron probe microanalysis (EPMA). Table 1 lists the composition and corresponding cutoff wavelength of InAsSb epilayers (Gao et al., 1999).

The growth process is as follows: Firstly, the InAs substrate and the source materials were set in the fused silica boat. The temperature of the furnace was increased to nearly 650°C and was kept at this temperature for 1 hour to mix the growth melt sufficiently. Next, the temperature was slowly decreased until 500°C with a cooling rate of 1°C/min. At this growth temperature, the melt contacts with the substrate, and the excess growth melt is immediately removed away from the substrate by pushing the melt holder. The key point is as follows: at the suitable growth temperature obtained by observation, some melt remains on the surface of substrate. Then the substrate was pushed under the flat part of the melt holder and cooled for 10°C at a cooling rate of 0.4°C/min to obtain an epilayer. Figs. 1(b)-(d) show the slideboat arrangement before contact, during contact with the melt, and during cooling for solidification, respectively. The thickness of the epilayer is dependent on the difference between the thickness of the substrate and the depth of the substrate well of the boat. Usually, the thickness of the grown layer reaches several decades ~ 100 μm. The samples were ground and polished using Al_2O_3 powder to obtain a flat and mirror-smooth surface for the characterization measurements.

We summarize the main differences between ME and LPE: (1) For the LPE, the growth process is controlled by solute diffusion. The solid composition is different from the melt composition. However, for the ME, the solute is not strongly distributed during the growth. The difference between the composition of the epilayer and that of the melt is small. (2) In the case of LPE, all of the growth melt is removed from the substrate after finishing growth. However, in the case of ME, a portion of the melt is aliquoted onto the substrate before growth, and this melt fraction is crystallized under the flat part of the melt holder.

x	Substrate	λ (μm)
0.946	InAs	9.5
0.948	InAs	10.5
0.95	InAs	11.0
0.96	InAs	12.5

Table 1. Compositional dependence of cutoff wavelengths of $InAs_{1-x}Sb_x$ epilayers (Gao et al., 1999)

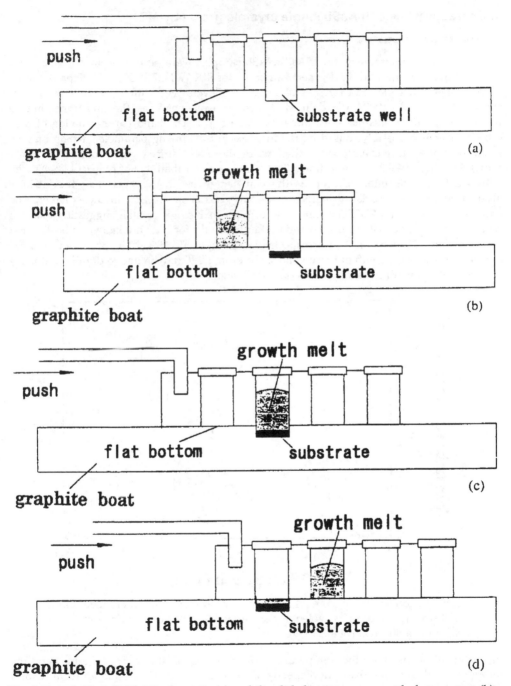

Fig. 1. Fused silica slideboat schematic (a) and the slide boat arrangement before contact (b), during contact with the melt (c) and during cooling for solidification (d) (Gao et al., 2002).

3. Characteristics of InAsSb single crystals grown by ME

3.1 Transmission spectra

Transmittance measurements for InAs/InAsSb samples were performed at 300 K using a Fourier transform infrared (FTIR) spectrophotometer (JIR-WINSPEC50, JEOL, Japan). In Fig. 2, the transmittance spectra for several InAsSb epilayers with different compositions are shown (Gao et al., 1999). We defined the cutoff wavelength as the mid-transmittance wavelength, and the results are summarized in Table 1. The band gap narrowing of ME-grown InAsSb is possibly caused by the increase in the bowing parameter, which may be induced by the lattice contraction of the ternary alloy. The lattice constant of $InAs_{0.051}Sb_{0.949}$ grown by ME is 6.4572 Å (see section 3.2), which is smaller than that of InSb (6.4789 Å). Fig. 2 shows that transmittances drop as the wavelength becomes longer than the intrinsic absorption region. This is attributed to the free carrier absorption in the epilayer. The absorption dip around 22-25 µm cannot be seen for InSb single crystals suggesting that it is related to InAs lattice vibrations. To and Lo Raman peaks for InAs are known to be 219 cm⁻¹ (45.6 µm) and 236 cm⁻¹ (42.4 µm) respectively (Gong et al., 1994). The observed dips are almost half the wavelength of these values, therefore they may be due to double frequency of lattice vibrations of InAs, i.e., the second harmonics.

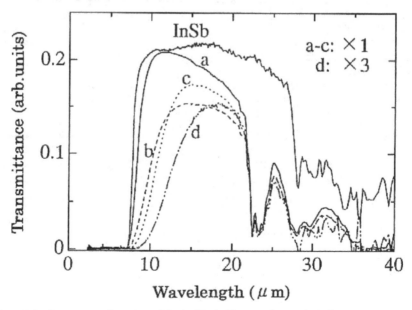

Fig. 2. Transmission spectra for several InAs/InAsSb samples with different compositions on InAs substrates. The cutoff wavelengths as defined by half transmittance are summarized in Table 1 (Gao et al., 1999).

In order to investigate the temperature dependence of the energy band gap, the temperature dependent absorption spectra were measured at 12, 77, 115 and 290 K for a $GaAs/InAs_{0.04}Sb_{0.96}$ sample with a total thickness of 600 µm. The energy band gaps of 0.147, 0.141, 0.136, and 0.112 eV were obtained respectively, at corresponding temperatures for this sample. To examine the temperature dependence of the energy band gap, we calculated the energy band gaps using a well-known Yarshni's empirical formula from 0 to 300 K (Yarshni, 1967):

$$E_g(T) = E_g(0) - \alpha T^2 / T + \beta \tag{1}$$

where $E_g(0)$ is the energy band gap at zero temperature, while α and β are two empirical parameters. Fig. 3 shows the calculated curve (solid line) and the date obtained from absorption measurements (opened squares) (Gao et al., 2006). Using least-square fits to equation (1), we obtained $\alpha = 0.147$ meV/K，$\beta = 60$ K，and E_g (0) = 0.1475 eV, respectively.

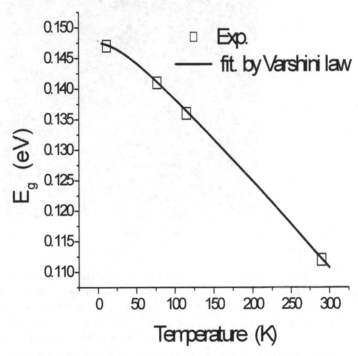

Fig. 3. Temperature dependence of energy band gaps for a GaAs/InAs$_{0.04}$Sb$_{0.96}$ sample (Gao et al., 2006).

3.2 Structural characteristics

The cross section of InAsSb epilayers was observed by scanning electron microscopy (SEM) (FEI Quanta 200F) at a magnification of 500. Fig. 4 shows a SEM cross-sectional image of an InAs/InAsSb sample. The boundary between the epilayer and substrate is flat and straight. It indicates the diffusion has not strongly occurred at the boundary for the growth temperature of about 500°C. This phenomenon may be benefitted from the special process of ME and the slow cooling rate of 0.4°C/min during the epitaxial growth. The cracks in the epilayer are caused by the cleavage. As is seen in Fig. 4, the thickness of the InAsSb epilayer grown by ME is about 40 μm, which is impossible to realize for MBE and MOCVD technologies. This thickness basically eliminates the effect of the lattice mismatch and results in the good crystal quality of the epilayer.

InAsSb epilayer ⟵

InAs substrate ⟵

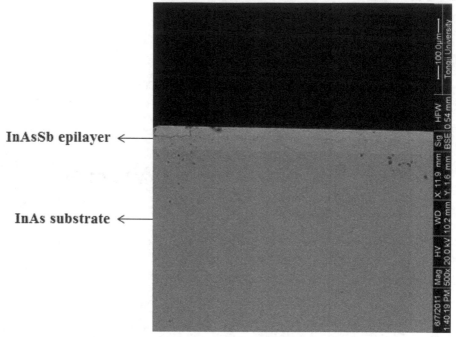

Fig. 4. Cross-sectional image of an InAs/InAsSb sample taken by SEM. The thickness of the InAsSb epilayer is about 40 μm.

Fig. 5 shows the X-ray diffraction (XRD) spectra of an InAs/InAsSb sample measured by an X-ray diffractometer (Rigaku D/MAX-2200PC, Cu barn) at a voltage of 40 kV, and a current of 40 mA (Gao et al., 2011). It is seen that (400) and (200) diffraction peaks of the InAsSb epilayer clearly appear, and no other crystal structures are observed. The growth direction of the epilayer is in agreement with the surface direction of the InAs substrate, i.e. the (100) orientation. This demonstrates that the InAsSb epilayer is indeed a single crystal. Since the thickness of the epilayer grown by ME reaches 40 μm, the diffraction peak of the InAs substrate does not appear. The sharpness and the full-width at half-maximum (FWHM) of 164 arcsec of the (400) diffraction peak indicate the high quality of InAsSb epilayers.

According to the Bragg diffraction equation, the lattice constant a for the InAs$_x$Sb$_{1-x}$ sample shown in Fig. 5 is estimated to be 6.4572 Å. Based on the Vegard Law, the InAs mole fraction in epilayers can be calculated as:

$$x = \left(a_{\text{InAsSb}} - a_{\text{InSb}}\right) / \left(a_{\text{InAs}} - a_{\text{InSb}}\right) \tag{2}$$

where x is the InAs mole fraction in the epilayers. The InAs mole fraction is calculated to be 0.051. The calculated lattice mismatch between the InAs$_{0.051}$Sb$_{0.949}$ epilayer and the InAs substrate is as large as 6.58%.

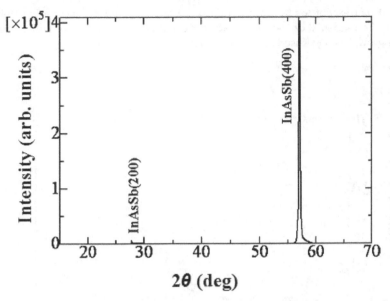

Fig. 5. XRD spectra of an InAs/InAsSb sample grown by ME. The lattice mismatch between the epilayer and substrate was estimated to be 6.58% (Gao et al., 2011).

3.3 Electrical properties

For measuring the electrical properties of the $InAs_{0.04}Sb_{0.96}$ epilayers, the surface of the samples should be ground and polished until mirror smooth, and the InAs substrate must be removed to eliminate the influence from the conductive InAs substrate. Moreover, an anodization treatment was performed on the surfaces of samples before measurements. Hence, the possibility of the presence of a surface conducting (accumulation) layer is also eliminated.

The carrier concentration and the electron mobility of InAsSb epilayers were measured using the Van der Pauw method with a standard Hall measurement system under a magnetic field of 2000 gauss. In was used as the contacts. To determine the temperature dependence of the electrical properties of the epilayers, the InAsSb sample holder was set in a cryostat. The temperature was changed from 300 to 50 K.

All $InAs_{0.04}Sb_{0.96}$ samples were n-type in the measurement temperature range. In Fig. 6 (Gao et al., 2004), measured carrier concentration as a function of temperature is shown for the two $InAs_{0.04}Sb_{0.96}$ epilayers with a thickness of 100 μm grown using a graphite boat and fused silica boat, respectively. It is seen that the carrier concentration of sample (a) grown using graphite boat has the same level of 0.2×10^{15} cm^{-3} at temperatures between 50 and 100 K, and then the carrier concentration gradually increases to 1.5×10^{15} cm^{-3} at 200 K. The carrier density of the $InAs_{0.04}Sb_{0.96}$ layer rapidly increases at the temperatures higher than 200 K due to the intrinsic generation of carriers in this temperature region. A carrier concentration of 2.3×10^{16} cm^{-3} was obtained at 300 K. This result is typical for narrow gap n-type InAsSb epilayers grown using a graphite boat by ME.

It is evident that except for the carrier density, which is about 3 times higher than that of sample (a), the tendency of the temperature dependence of the carrier concentrations for sample (b) grown using a fused silica boat is similar to that of sample (a).

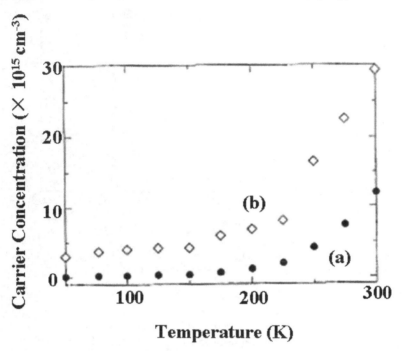

Fig. 6. Measured temperature dependence of carrier concentration for the two InAs/InAs$_{0.04}$Sb$_{0.96}$ epilayers grown using (a) graphite boat; (b) fused silica boat (Gao et al., 2004).

Fig. 7 shows the measured temperature dependence of the electron mobility for the two InAs/InAs$_{0.04}$Sb$_{0.96}$ epilayers with a cutoff wavelength of 12 μm grown using graphite and fused silica boat (Gao et al., 2004).

It is seen that the electron mobility of sample (a) is lower than 2×10^4 cm^2/Vs at temperatures below 100 K, then quickly increases until 200 K. Electron mobility decreases again as temperature continuously increases to 300 K. Peak electron mobilities of 1×10^5 cm^2/Vs at 200 K, and 6×10^4 cm^2/Vs at 300 K have been obtained. This is the best result so far for InAsSb materials with cutoff wavelengths of 8-12 μm indicating the good quality of epilayers. Polar optical phonon scattering governs electron mobility at high temperatures. At low temperatures, alloy scattering is not important in these InAs$_{0.04}$Sb$_{0.96}$ epilayers because the As composition is only 0.04, and a dislocation density of the order of 10^4 cm^{-2} has been observed by counting the dislocation amount on the surface of the epilayers after chemical etching. Therefore, the rapid decrease in measured electron mobility with decreasing temperature below 175 K is probably due to impurity scattering.

The electron mobilities of sample (b) are all in the range of 4-5.27×10^4 cm^2/Vs at temperatures between 50 and 300 K. Only the peak points at 50 K and 225 K show mobilities higher than 5×10^4 cm²/Vs. The most important phenomenon is that the electron mobility at 77 K is higher than that at 300 K. The purity improvement has possibly resulted from the reduction of carbon contamination in fused silica boat. This gives rise to a weaker impurity scattering in InAsSb epilayers grown with fused silica at 77 K.

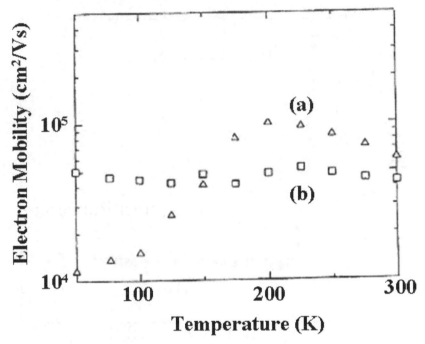

Fig. 7. Measured electron mobility as a function of temperature for the two InAs/InAs$_{0.04}$Sb$_{0.96}$ samples grown using (a) graphite boat; (b) fused silica boat (Gao et al., 2004).

The relation of ln n and T^{-1} of a GaAs/InAs$_{0.04}$Sb$_{0.96}$ sample is shown in Fig. 8, where ln n is the natural logarithm of electron concentration, and T^{-1} is the inverse of the temperature. As shown in Fig. 8, the saturation range appears between 40 K and 200 K. In this temperature range, donors are mainly ionized, i.e. most of electrons in donor level are excited to conduction band. Thus Fermi level is lower than donor level. However, most of electrons of valence band have not yet been excited to conduction band in this temperature region. When temperature increased to 200-300 K, the intrinsic region is observed as seen from Fig. 8. The electrons excited from valence band become the dominant conductive electrons due to intrinsic transition of carriers. Therefore, GaAs/InAs$_{0.04}$Sb$_{0.96}$ samples exhibit the intrinsic property in this temperature region. We obtained the relation between ln n and T^{-1} in the intrinsic region as follows (Yishida et al., 1989):

$$\ln n = (-E_g / 2k)T^{-1} + \ln(N_C N_V)^{1/2} \tag{3}$$

where N_C is effective density of state of conduction band, N_V is effective density of state of valence band, E_g is energy band gap, and k is Boltzmann's constant. In terms of equation (3), we calculated energy band gap E_g from the slope of intrinsic region in Fig. 8. The calculated energy band gap is 0.1055 eV, which is well consistent with the result obtained from room temperature transmittance measurements (0.1033 eV). This result demonstrates that the energy band gap of GaAs/InAs$_{0.04}$Sb$_{0.96}$ alloys grown by ME is strongly narrowed.

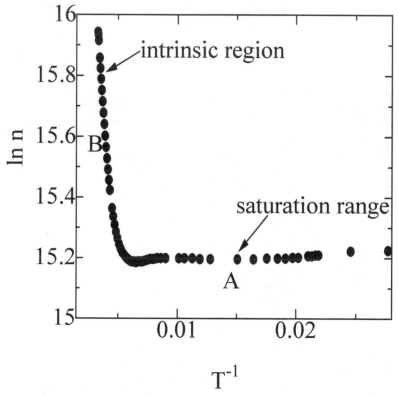

Fig. 8. Relation of ln n and T^{-1} for a GaAs/InAs$_{0.04}$Sb$_{0.96}$ sample (Gao et al., 2006).

4. Uncooled InAsSb photoconductors with long wavelength

4.1 Fabrication of immersion detectors

The immersion photoconductors were fabricated using InAs/InAsSb materials grown by ME. Since the thickness of the epilayers reached several decades ~ 100 μm, InAs substrates can be entirely removed away. Thus the effect of the lattice mismatch between epilayers and substrates were eliminated. In the device process, InAsSb epilayers were thinned to about 15 μm by grinding. Chemical polishing was carried out using a $C_3H_6O_3$ etchant after grinding to eliminate the mechanical damage to the wafers. Indium was used as the ohmic contact. The light-sensitive area of the photoconductors is 0.05 × 0.05 cm^2. The detector resistances are 20-110 Ω at 300 K. Ge optical lenses were set on the detectors. There are not any antireflective coatings deposited on the surfaces of the elements and lenses. Fig. 9 shows a photograph of uncooled InAsSb photoconductors.

Fig. 9. Photograph of uncooled InAsSb photoconductors.

4.2 Spectral photoresponse

The spectral photoresponse of InAsSb photoconductors were measured by a Fourier transform infrared (FTIR) spectrometer at room temperature, and the absolute responsivity was calibrated by a standard blackbody source at a temperature of 500 K and a modulation frequency of 1200 Hz. The bias current applied on the devices was 10 mA. Fig. 10 shows the detectivity D^* varying with the wavelength of uncooled $InAs_{0.051}Sb_{0.949}$ immersion photoconductors. At room temperature, the peak voltage responsivity is 164.3 V/W at the wavelength of 6.5 μm, resulting in the corresponding peak detectivity as high as 2.5×10^9 cm $Hz^{1/2}$ W^{-1}. The peak detectivity of our detectors is more than one order of magnitude higher than that of long-wavelength type-II InAs-GaSb superlattice photodiodes (Mohseni et al., 2001). This may be caused by the following reasons: (1) the super-hemisphere immersion Ge lenses were set on our photoconductors. The incident IR radiation was focused by the lenses, thus the radiation energy density on the photosensitive surfaces was raised. It is well known that the super-hemisphere immersion component is able to increase the signal-to-noise ratio and detectivity of n_s^2 multiples, $n_s = 4$ is the refractive coefficient of Ge crystals. (2) ME-grown InAsSb epilayers with the thickness of several decades ~ 100 μm are the narrow gap materials, and have the properties of bulk single crystals. The narrow gap intrinsic semiconductors are particularly suitable for room temperature photon detector fabrication. The high density of states in the valence and conduction bands of them leads to the strong absorption of IR radiation (Piotrowski et al., 2004). As shown in Fig. 10 that D^* of 8.8×10^8 and 6.3×10^7 cm $Hz^{1/2}$ W^{-1} at the wavelength of 8.0 and 9.0 μm respectively, have been obtained.

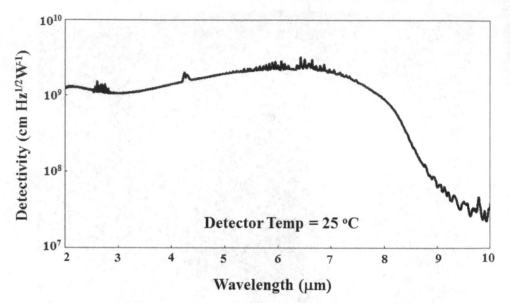

Fig. 10. Detectivity D^* versus wavelength of InAs$_{0.051}$Sb$_{0.949}$ immersion photoconductors at room temperature.

5. Conclusion

We provided high sensitivity uncooled InAsSb photoconductors with long wavelength. Ge optical lenses were set on the photoconductors without any antireflective coatings. The detectors are based on InAsSb epilayers with the thickness of 40 μm grown on InAs substrates by melt epitaxy. This thickness efficiently eliminates the effect of lattice mismatch and results in a low dislocation density (the order of 10^4 cm^{-2}) in epilayers, thus improves the terminal performance of detectors.

FTIR transmission spectra for InAs$_{1-x}$Sb$_x$ epilayers revealed the strongly band gap narrowing. The temperature dependence of energy band gap for GaAs/InAs$_{0.04}$Sb$_{0.96}$ was studied between 12 K and 290 K by measuring the absorption spectra. The structural properties of InAsSb materials were characterized by SEM observation and XRD spectroscopy. The measurement results of the materials exhibited the high quality.

The electrical properties were investigated by van der Pauw measurements. A peak electron mobility of 100,000 cm^2/Vs with a carrier density of 1×10^{15} cm^{-3} at 200 K, and an electron mobility of 60,000 cm^2/Vs with a carrier density of 2.3×10^{16} cm^{-3} at 300 K, have been obtained for an InAs$_{0.04}$Sb$_{0.96}$ epilayer. This is the best result so far, to our knowledge, for the InAsSb materials with cutoff wavelengths of 8-12 μm indicating the good quality of the epilayers. The purity of the epilayers grown with fused silica boat was improved. The purity improvement has possibly resulted from the reduction of carbon contamination in fused silica boat. A room temperature band gap of 0.1055 eV is demonstrated via analyzing the temperature dependence of the carrier density for the GaAs/InAs$_{0.04}$Sb$_{0.96}$ layers, which is in good agreement with the value obtained by transmittance measurements.

The spectral photoresponse of InAsSb photoconductors grown by ME were measured using a FTIR spectrometer at room temperature. The bias current applied on the devices was 10 mA. At room temperature, the peak detectivity $D_{\lambda p}*$ (6.5 μm, 1200) reaches 2.5×10^9 cm $Hz^{1/2}$ W^{-1} for InAsSb immersion photoconductors. The detectivity $D*$ at the wavelength of 8 μm is 8.8×10^8 cm $Hz^{1/2}$ W^{-1}, and that at 9 μm is 6.3×10^7 cm $Hz^{1/2}$ W^{-1}. The excellent performance of the detectors indicates the potential applications for IR detection and imaging.

6. References

[1] J. D. Kim, D. Wu, J. Wojkowski, J. Piotrowski, J. Xu, and M. Razeghi, "Long-wavelength InAsSb photoconductors operated at near room temperatures (200-300 K)", Appl. Phys. Lett., 68, No. 1, 99 - 101 (1996).

[2] A. Kumar and P. S. Dutta, "Growth of long wavelength $In_xGa_{1-x}As_ySb_{1-y}$ layers on GaAs from liquid phase", Appl. Phys. Lett., 89, No. 16, 162101-1 - 162101-3 (2006).

[3] J. -I. Chyi, S. Kalem, N. S. Kumar, C. W. Litton, and H. Morkoc, "Growth of InSb and $InAs_{1-x}Sb_x$ on GaAs by molecular beam epitaxy", Appl. Phys. Lett., 53, No. 12, 1092 - 1094 (1988).

[4] V. K. Dixit, B. Bansal, V. Venkataraman, H. L. Bhat, K. S. Chandrasekharan, and B. M. Arora, "Studies on high resolution x-ray diffraction, optical and transport properties of $InAs_xSb_{1-x}$/GaAs ($x \leq 0.06$) heterostructure grown using liquid phase epitaxy", J. Appl. Phys., 96, No. 9, 4989 - 4997 (2004).

[5] Y. Z. Gao, H. Kan, F. S. Gao, X. Y. Gong, and T. Yamaguchi, "Improved purity of long-wavelength InAsSb epilayers grown by melt epitaxy in fused silica boats", J. Cryst. Growth., 234, 85 - 90 (2002).

[6] Y. Z. Gao, X. Y. Gong, Y. H. Chen, and T. Yamaguchi, "High Quality $InAs_{0.04}Sb_{0.96}$/GaAs Single Crystals with a Cutoff Wavelength of 12 μm Grown by Melt Epitaxy", Proc. of SPIE., 6029, 60291I-1 - 60291I-7 (2006).

[7] Y. Z. Gao, X. Y. Gong, G. H. Wu, Y. B. Feng, T. Makino, and H. Kan, "Uncooled InAsSb Photoconductors with Long Wavelength", Jpn. J. Appl. Phys., 50, No. 6, 060206-1 - 060206-3 (2011).

[8] Y. Z. Gao, X. Y. Gong, H. Kan, M. Aoyama, and T. Yamaguchi, "$InAs_{1-y}Sb_y$ Single Crystals with Cutoff Wavelength of 8-12 μm Grown by a New Method", Jpn. J. Appl. Phys., 38, No. 4A, 1939 – 1940 (1999).

[9] X. Y. Gong, H Kan, T. Yamaguchi, I. Suzuki, M. Aoyama, M. Kumagawa, N. L. Rowell, A. Wang, and R. Rinfret, "Optical Properties of High-Quality $Ga_{1-x}In_xAs_{1-y}Sb_y$/InAs Grown by Liquid-Phase Epitaxy", Jpn. J. Appl. Phys., 33, No. 4A, 1740-1746 (1994).

[10] Y. P. Yarshni, "Temperature dependence of the energy gap in semiconductors", Physica A., 34, 149 (1967).

[11] Y. Z. Gao, X. Y. Gong, Y. S. Gui, T. Yamaguchi, and N. Dai, "Electrical Properties of Melt-Epitaxy-Grown $InAs_{0.04}Sb_{0.96}$ Layers with Cutoff Wavelength of 12 μm", Jpn. J. Appl. Phys., 43, No. 3, 1051 - 1054 (2004).

[12] T. Yishida and A. Shimizu, Semiconductor Devices, Corona Publishing Co., LTD, Tokyo, 1989. (In Japanese)

[13] H. Mohseni and M. Razeghi, "Long-Wavelength Type-II Photodiodes Operating at Room Temperature", IEEE PHOTONICS TECHNOLOGY LETTERS, 13, No. 5, 517 - 519 (2001).

[14] J. Piotrowski and A. Rogalski, "Uncooled long wavelength infrared photon detectors", *Infrared Phys. Technol.*, 46, 115 - 131 (2004).

5

Terahertz Emitters, Detectors and Sensors: Current Status and Future Prospects

M. Ghanashyam Krishna[1,2], Sachin D. Kshirsagar[1] and Surya P. Tewari[1,2]
[1]Advanced Centre of Research in High Energy Materials
[2]School of Physics, University of Hyderabad, Hyderabad
India

1. Introduction

The region of the electromagnetic spectrum from 0.3 to 20 THz (10 – 600 cm[-1], 1 mm – 15 µm wavelength) is an area for research that encompasses physics, chemistry, biology, materials science and medicine. The lack of sources of high quality radiation has limited developments in this area. However, there has been an upsurge of activity in the last five years due to the emergence of a wide range of new technologies. Terahertz radiation is now available in both continuous wave (cw) and pulsed form, down to single-cycles or less, with peak powers up to 10 MW [Ginzburg et al., 2011; Sizov& Rogalski, 2010; Williams, 2004; Tochitsky et al., 2005]. There is a wealth of information that can be extracted from a study of matter (solid, liquid or gaseous) at these frequencies. It is becoming evident that such directed basic research would also lead to a phenomenally large number of applications covering Physics, Chemistry, Life Sciences and Engineering [Jacobsen et al., 1996; Mueller, 2003; Chamberlain, 2004; Bolivar et al., 2004; Lewis, 2007; Tonouchi, 2007] .

As stated earlier, to achieve further understanding of physical phenomena in the THz region of the electromagnetic spectrum it is essential that there are reliable sources (including materials, devices and integrated systems) that emit THz radiation. Furthermore, there is need for materials, devices and systems that detect this radiation. Another aspect, which is central to all developments in THz science and technology is the development of reliable standards. Thus, THz metrology is another area that has to develop simultaneously with development of emitters and detectors. One of the applications, which is still in its infancy but has remarkable potential, is THz sensors for a variety of contexts in security, life sciences and medicine.

The aim of this review is to present the state-of-science and technology in the area of THz emitters, detectors and sensors. A simple search carried out on a general search engine with keywords "THz detectors or THz emitters" comes up with thousands of hits. Even a search on more focused science literature search engines and limited to the last five years, with the same keywords, generates over a thousand papers. This in essence, reveals the explosion of interest in the area of THz science as well as the ease of availability of sources, probes and tools for such studies. Therefore, the current review is limited to, what the authors consider, the most interesting and significant recent developments in the area of THz emitters, detectors and sensors. It is aimed at researchers intending to initiate work in these fields and therefore a considerable portion is focussed on basic principles. The rest of the review is organized in the following fashion: Section 2 will review some of the fundamental phenomena that occur in the THz region of the em spectrum; Section 3 will present a review on THz emitters; Section 4 is focused on THz detectors and sensors and; Section 5 will present an overview of the materials development work being carried out by the present authors. Section 6 will provide a summary and outlook for the future in this field

2. Significance of THz

The THz spectral range is characterized by several unique features [Jacobsen et al., 1996; Mueller, 2003; Chamberlain, 2004; Bolivar et al., 2004; Lewis, 2007; . Innocenzi et al., 2009; Roeser et al. 2010]. It is the highest frequency band where the field can be measured coherently without an interferometer. The THz region of the spectrum is sensitive to all of the thermally accessible excitations that determine the properties of correlated electron systems. Also, the ability to perform spectroscopy in the time as well as frequency-domain provides a unique window on the temporal evolution of optical response functions on time scales between 100 fs and 500 ps. Furthermore, THz methods are compatible with the study of materials under extreme conditions of temperature, electric, and magnetic field. THz probes are, in general, non-contact enabling polarization analysis and polarization control that may be of importance in situations where Ohmic contacts to certain materials are not possible due to the remote location of the material, or to avoid the introduction of extrinsic effects associated with leads. Terahertz probes can also provide detailed information on interface quality or on the presence of buried defects or structures.

Two, one and zero dimensionally quantum confined systems have electronic excitations in the THz regime. As the THz frequency exceeds various relaxation rates and broadening, it is possible to achieve well defined collective and single particle excitations. As a consequence many fundamental issues relating to control of carrier injection by doping, the relationship between geometry and energy level structure, line widths of transitions between states, excited state coherence and population lifetimes, superposition states of carriers, and spin excitations can be probed. These are issues not only of interest from a basic perspective but also critical for applications such as sources, detectors, novel ultrafast electro-optic devices and quantum information processing in semiconductors [Tonouchi, 2007].

Another area of interest is the coherent quantum control of the orbital and spin states of carriers in semiconductor nanostructures that occur at THz frequencies. The properties of semiconductor nanostructures at visible and near-IR wavelengths can also be controlled with THz-frequency electromagnetic radiation. In these structures, energy-level splittings are often in the THz range. At THz frequencies, electron interactions with acoustic and optical phonons are small leading to narrow line widths of electronic excitations. These small line widths and correspondingly long relaxation and dephasing times find application in a variety of devices, including light emitters, detector and quantum logic devices Magnetic level splittings in semiconductor nanostructures in magnetic fields of a few Tesla are frequently in the THz regime. Metals are typically very good reflectors of THz radiation. Majority of solids have optical phonons at THz frequencies. Most of these are infrared-active, which can be excited directly by resonant oscillating electric fields[Lewis, 2007]. Phonon features are particularly interesting near structural phase transitions, where they become "soft" and move towards zero frequency. Phonons interact strongly with electrons, a phenomenon that was central to explaining conventional superconductivity. A Josephson plasma resonance is produced by coherent propagation of Cooper pairs between superconducting "sheets" in layered superconductors. In a variety of layered superconductors this mode falls in THz frequency range[Tonouchi, 2007; Ferguson & Zhang, 2002; Sizov & Rogalski, 2010] Other similar strongly correlated electron systems have very complex phase diagrams with several transitions among them. Understanding the nature of these phases, as well as the transitions among them, requires detailed characterization of their excited states, or elementary excitations. The most important excitations are those with an energy of order $k_B T$ above the ground state and therefore in the THz region of the electromagnetic spectrum. Fundamental interactions between quasiparticles, phonons, spin-excitations and other constituents of correlated materials occur on ultrafast time scales [Kida et al., 2005]. It is worth noting that the phonon modes of inorganic and organic crystals fall in the THz region of the spectrum [Zheng et al., 2007].

The classical rotation period for a nitrogen molecule with one $\hbar$ of angular momentum is about a pico second. Similarly, the time of a molecular collision at room temperature is 0.1 to 1 ps. At very low temperatures the collisional energy of molecular collisions is comparable to the rotational energy level spacing. As a consequence, the collisions are quantum mechanical in nature and resonances in their cross sections occur in the THz region. The time scale of gas phase collisions is of the order 10^{-12} seconds which is ideally suited for THz studies. Rydberg atoms in the $n = 20$ to $n = 60$ range span the spectrum of sensitive response to THz fields. These are model systems for studying many aspects of few body quantum mechanics, including quantum chaos, quantum-classical correspondence, and quantum systems in the presence of strong external fields [Rangan & Bucksbaum, 2001] .

Interaction of high-intensity femtosecond lasers with solid matter generates plasmas that can be employed as sources of short x-ray pulses, coherent harmonic radiation, energetic electron and ion beams, plasma-based accelerators, ultra high magnetic fields. These plasmas can also serve as a source of intense THz emission [Li et al, 2011].

There are several processes in liquids that occur in the THz region and lend themselves to be probed using THz radiation.THz spectroscopy can also reveal information about protein structure and dynamics. It is possible to distinguish many amino acids based on their THz spectrum, particularly in the crystalline form. The THz spectra of individual DNA base pairs can be used to understand dynamics and conductivity. It could lead to label-free measurement of protein-protein interactions as cellular activity is occurring in live cells. THz radiation holds promise in new medical imaging techniques based on pulsed and cw THz sources. Differences in tissue water, architecture and chemical content can easily be detected using THz techniques since this radiation is strongly absorbed by water. This can lead to early detection of diseases by revealing features that are not apparent with other imaging techniques. [Nagel etal., 2003] .

It is evident that THz region is a region of the electromagnetic spectrum that is rich with information on all forms of matter. It is further obvious that significant developments in this area are technology limited rather than being limited by science.

The research and development in THz science and technology can be broadly divided in to the following categories based on the fundamental phenomena that occur at the THz region of the em spectrum. In all these cases there are issues relating to materials, processing technologies, devices, and system integration issues need to be addressed.

1. Sources
2. Emitters
3. Detectors and sensors

It is pertinent at this point to distinguish between detectors and sensors. For purposes of this review, detectors are used to signify components that provide only frequency information of the THz radiation. Sensors, in contrast, provide information on both the frequency and time domain as well as the source that causes variation in frequency of radiation incident on the detector. Detectors, thus, form a sub-set of the sensor. Sensors can, therefore be used for imaging, spectroscopy and sensing the nature of environment around the THz emitter or detector.

3. THz emitters

Terahertz radiation has potential in various short-wavelength communication devices and security applications. In order to attain the potential offered by terahertz technology, design of solid state terahertz emitters is indispensible. Terahertz sources based on femto second lasers pulses is one of the most promising techniques to generate THz radiation. Lasers are strong sources of electromagnetic energy where energy is stored both in the intensity and frequency. THz generation by ultrafast laser pulses can be classified as linear and non linear optical processes. The linear process involves injection of photocarriers by a laser pulse into the semiconductor connected to the antenna. The photocarriers are accelerated by dc bias applied to the antenna. The antenna radiates THz signal of a frequency that is determined by the pulse duration of laser pulses. In contrast, in the nonlinear optical process THz radiation is generated by optical rectification or difference frequency mixing techniques. The nonlinear optics methods are attractive for terahertz devices because of several properties, including (i) possibility of room-temperature operation (ii) easy operation. However, one of the major hurdles in operating terahertz emitters lies in poor power conversion efficiency, which is of the order of 10^{-6} to 10^{-5}. Terahertz sources can be divided according to their bandwidth, output power, and operating principle [Sakai & Tani, 2005]. In following sections we focus on the several kinds of terahertz emitters. Brief theories of the operating principle have been put forth with the explanation of their merits and demerits.

3.1 Photoconductive broadband THz antenna

An ultrashort laser pulse incident on the surface of a photoconductor at an angle of incidence θ, generates photocarriers, which then move under the influence of an electrical bias field. The bias field may be either externally applied (parallel or perpendicular to the surface) or internally generated by band bending from trapped surface charges (perpendicular to the surface). The resulting transient photocurrent gives rise to the emission of THz electromagnetic pulses in the reflected and transmitted directions. When a femto second (fs) laser pulse, with an arbitrary intensity profile, excites a biased semiconductor with photon energy greater than its bandgap, electrons and holes are produced at the illumination point in conduction and valence bands, respectively. The rapid changes of the density of photocarriers and their acceleration due to the applied DC bias produce an electromagnetic field radiation into free-space with the help of an antenna. The production of ultrashort currents with a FWHM lifetime of 1 ps or less strongly depends on the carrier's lifetime in the semiconductor. The space-charge screening and radiation-field screening are two factors preventing photoconductor from generating higher power THz radiation, which could affect all kinds of the photoconductive antenna. Based on the shape of antenna, their effects are different. Generally speaking, the space-charge screening can be important to small dipole antennas, but, for large area antennas, radiation-field screening is the major cause of saturation. Photoconducting switches are commonly used in time-domain terahertz spectroscopy, where, a generation and detection THz wave is carried out by use of biased photoconducting GaAs aperture antenna. The advantage of semiconductor photoconductive switches is that they can be used to efficiently generate signals and to generate and detect electrical transients in guided media or free space. However, the problem with photoconductive switches is that electrical pulse width is limited by a number of factors such as the laser pulse width, circuit parameters of the generation and detection site, and the carrier lifetime in the semiconductor. Furthermore, the carriers can be captured by traps at the grain boundaries, resulting in a rapid decay in number of carriers [Sizov & Rogalski, 2010].

Usually materials with reduced carrier lifetimes and high carrier mobility such as LT-GaAs or ion implanted semiconducting layers are utilized. Ga doped InAs based photo conductive switches are expected to yield higher terahertz output because of their higher electron mobility [Lloyd-Hughes et al., 2005]. Liu *et al.* report improvement in photoconductive response in shorter terahertz emission pulses from spiral antennas fabricated on multi-energy-implanted GaAs:As[Liu et al., 2005]. This is due to the uniformity of arsenic antisite defects (traps) distributed in multi-energy-implanted GaAs:As which contributes to the unchanged carrier trapping time as compared to single-energy-implanted GaAs:As. Miyadera *et al.* [Miyadera et al., 2004] used amorphous Ge thin films fabricated on (1 0 0) MgO substrates as a terahertz emitter by coupling with Au/Cr transmission lines. Amorphous Ge will be beneficial for optical to electrical signal converter, due to the potential advantages of fabrication without heating process and smaller bandgap of about 0.66 eV.

Light absorption process in the photoconductive switches which generates free carriers near the surface must also be considered. The absorption of terahertz pulses by free carrier can degrade overall efficiency of photoconductors. GaAs constituents of photoconductive

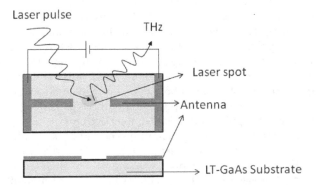

Fig. 1. Typical schematic of a photoconductive switch

switches have a phonon resonance around 8 THz that generates a broad absorption band between 7 and 10 THz. In order to reduce absorption caused by phonon resonance, Kasai *et al.* have fabricated LT-GaAs epitaxial layer which is transferred onto high resistivity Si substrates [Kasai et al., 2009]. These detectors showed higher sensitivity than conventional photoconducting switches at high frequencies. The typical schematic of a photoconductive switch is shown in fig 1.

3.1.1 Photo-Dember effect emitter

If surface energy bands of a semiconductor lie within its bulk band gap, then Fermi level pinning occurs, leading to band bending and the formation of a depletion/accumulation region at the surface. The resulting electric field will separate photo generated electrons and holes, forming a dipole perpendicular to the surface which emits THz radiation, this is the surface-field emission. By changing doping from n type to p type, the sign of the electric-field direction is reversed. This leads to a change in the polarity of the dipole, which is observed as a change in sign of the generated THz transient. A photo-Dember field can occur at the surface of a semiconductor after photo excitation. Two factors lead to this field: a difference in diffusion coefficients for electrons and holes, and a structural asymmetry. In a typical semiconductor, electrons have a larger diffusion coefficient than holes. Therefore, after photo excitation, the electron population diffuses more rapidly than the hole population. In the absence of a surface boundary, there would be no net dipole field, since the center of charge does not change. However, in the vicinity of the surface, reflection or capture of charges results in the center of charge for electrons and holes moving away from the surface. A dipole is thus formed perpendicular to the surface, leading to THz emission. In this case changing the semiconductor doping from n to p type has no effect on the sign of the emitted THz pulse. One of the main applications of photo-Dember effect is the generation of terahertz (THz) radiation pulses for terahertz time-domain spectroscopy. This effect is present in most semiconductors but it is particularly strong in narrow-gap semiconductors (mainly arsenides and antimonides) such as InAs and InSb owing to their high electron mobility.[Yi et al, 2010; Reklaitis, 2011; Klatt et al., 2010; Krotkus, 2010]

3.1.2 Current transient effect emitter

The current transient emitter or Auston switch is simply a gap in a uniform transmission line defined by thin-film metal fabrication on the top surface of an ultrafast photoconductor The transmission line is, generally, a coupled-strip line terminated in a planar antenna, within a planar strip dipole. The entire transmission line structure is biased with a voltage through an inductor. The principle of operation is that the gap in the transmission line initially creates an open-circuit condition. When a short pulse from a mode-locked laser arrives in the gap, a shower of electron-hole pairs are generated in the first few microns below the surface. The instantaneous concentration of these pairs is high enough to "short-out" the gap. When the laser pulse is terminated, the electron-hole pairs quickly recombine, creating an electrical impulse (or surge current) on the same time scale as the laser pulse. Due to the wide bandwidth of the planar transmission line, the electrical impulse propagates down the line to the antenna with minimal dispersion or droop. So upon reaching the antenna, the electrical pulse generates a significant amount of THz radiation that propagates primarily into the substrate.[Reimann, 2007; Suen et al., 2010;]

3.2 Optical rectification based emitters

One of the more popular means of generating THz frequency radiation is through the bulk second-order nonlinear effect of optical rectification. Optical rectification occurs if high-intensity light is directed onto an electro-optic material. If the excitation beam contains components of two or more frequencies then difference-frequency mixing, also known as optical rectification, may take place. Depending on the spectrum of the excitation beam, the resulting frequency may be in the THz range. The optical rectification process has been widely developed for THz generation, typically in materials with a completely nonresonant response. The advantage of such an approach is the inherent speed of the response, which may be viewed as essentially instantaneous. The disadvantage is the relative weakness of the nonlinear response. Thus, in order to obtain significant THz emission via the optical rectification mechanism it is desirable to use a medium of appreciable thickness which leads to the necessity of considering phase-matching constraints. The problem of broadband phase matching is a challenge. One possibility is exploiting the non-resonant second-order nonlinear response expected in any non-centrosymmetric semiconductor, such as the III–V materials. This effect can be identified by the distinctive dependence that it shows on the crystallographic orientation of the sample. Optical rectification using an ultra-short femto laser in the periodically poled structure can be considered as a special case of difference frequency mixing where terahertz radiation emitted from the polarization change that follows the transport of excited carriers in an applied or surface electric field. The physics of optical rectification leading to THz generation is well understood [Blanchard et. al, 2011]. The nonlinear coefficient of the second order susceptibility is an extremely important material property when high-power output THz emitters are considered. The output power is also predominantly determined by intensity of the excitation pulse and the phase matching conditions. The major disadvantage of this technique is low power output, broad line-width, velocity mismatch and requirement of expensive femto second laser sources. In addition, there are the restrictions implied by other nonlinear processes like two-photon absorption, competing with optical rectification and limiting its efficiency. The process of two-photon absorption not only influences efficiency of terahertz generation but also impacts the terahertz spectrum [Vidal et al., 2011]. THz generation via optical rectification of x mode laser in a rippled density magnetized plasma has been demonstrated [Bhasin and Tripathi [Bhasin & Tripathi, 2009]. Fiber lasers are attractive as terahertz emitters because of their low cost, small size, high repetition rate, high pulse energy, and short pulse duration[Hoffmann et al., 2008].

Relativistic optical rectification, has opened up the potential realization of high frequency range terahertz emitter[Tsaur & Wang, 2009]. It is should be noted that bandwidth of emitter is sharply reduced in optical rectification due to the phase-matching condition. For EO sampling and optical rectification, the spectral response of the material is strongly sensitive to reststrahlen band absorption/reflection, phase matching, dispersion of refractive index and dispersion of EO coefficients, which can be expressed as

$$G(\Omega) = \frac{t(\Omega)(\exp(i2\pi\Omega\,\delta(\Omega)) - 1)}{\left[1 - r(\Omega)^2 \exp(\frac{i4\pi\Omega n(\Omega)d}{c})(i2\pi\Omega\delta(\Omega))\right]} \tag{1}$$

$$\delta(\Omega) = \frac{n_g(\lambda) - n(\Omega)}{c} d \tag{2}$$

$\delta(\Omega)$ is the velocity mismatch in time, $n_g(\lambda)$, is the optical group index, $n(\Omega)$, is the terahertz refractive index, d is crystal thickness, $t(\Omega)$ and $r(\Omega)$ are the Fresnel transmission and reflection coefficients and the multi-reflection of terahertz field inside the sensor material. The velocity mismatch between phase velocity of terahertz pulse and group velocity of laser beam hampers output power of these sources. This could be eliminated by placing nonlinear crystal in hollow metallic waveguide structure [Nikoghosyan, 2010]. The conversion efficiency in optical rectification depends primarily on the material nonlinear coefficient and the phase-matching conditions and various schemes for power enhancement have been reported [Radhanpura et al., 2010; Bugay & Sazonov, 2010; He et al., 2008; Reid et al., 2008]. To overcome the problem of phase mismatch, interaction of pump optical pulse with a two-level resonant impurities, for the terahertz wave is proposed [Bugay & Sazonov, 2010]. Alternatively, phase mismatching can also be effectively used to tune the operating frequency from 4.2 to 1.1 THz of a (110) oriented ZnTe crystal terahertz emitter. Materials with large drift velocity and short carrier lifetime such as GaAs and ZnTe are popularly used in optical rectification techniques. However, they are limited by large band gap which hampers their efficiency [Li & Ma, 2008]. In combination with second order susceptibility, effective second-order processes such as electric field induced optical rectification (EFIOR) has been shown to yield terahertz generation in centrosymmetric Ge crystals [Urbanowicz et al., 2007]. This effect has been termed EFIOR, because it is easily explained in the framework of third-order nonlinearity ($\chi^{(3)}$) contributed to effective second order process by DC surface electric field.

3.3 Quantum cascade laser emitters

Energy levels in semiconductor quantum wells can be designed and engineered to be of any value, with ease. In such systems, inter sub band transitions involving electrons that make lasing transitions between subband levels within the conduction band can be exploited to realize emissions in the far infrared and THz regions. Such lasers are called quantum cascade lasers (QCLs). Simple calculations show that energy level separation to achieve emission in the THz region from QCLs is in the meV region (1 THz = 4 meV). Design of quantum well structures for selective injection to the upper level and selective depopulation of electrons from the lower level is a challenge since, due to the narrow separation between subband levels, heating and electron-electron scattering have a great effect. Some of the other challenges that need to be overcome include detection and analysis of spontaneous emission; mode confinement at longer wavelengths; dielectric waveguide confinement (as the evanescent field penetration, which is proportional to the wavelength is of the order of several tens of microns>> the active gain medium of several microns. The physical origin of the THz emission of the quantum cascade laser, as stated above, can be traced to intersubband transitions. The quantum well thickness in QCL is one of the key parameters to control in order to finely tune laser energy. The background blackbody radiation and/or emissions from impurity levels may lead to poor reproducibility of the results. The micro-probe photoluminescence has been shown to provide a powerful tool for this purpose which can probe emission from individual states in the active region with the high spatial resolution [Freeman et al., 2011].

Unfortunately, it has been observed that QCL are not popular for THz source due to low energy (long wavelength) output. This is a consequence of the poor coupling between the small gain medium and the optical field. In fact, high optical losses are caused due to free electrons in the material at low energy. Recent developments in QCL technology have raised the hope for their application in terahertz devices. The nonequilibrium Green's function calculations performed by Yasuda et al. [Yasuda et al., 2009] predict that the 4L QCL has a larger terahertz gain than the conventional resonant phonon QCL. They have attributed it to the large number of electrons accumulate in the upper lasing level and contribute to lasing in the new scheme. They have found that the advantage of gain deteriorates at 200 K due to thermally activated phonon scattering.

Another major limitation of QCL is their low operation temperature. The maximum operating temperature of 178 K is currently the highest operating temperature achieved in a THz QCL. The need to use cryogenic cooling is considered to be a major obstacle to the introduction of room temperature QCL in terahertz devices. Ambient temperature operation can be achieved by integrating the optical nonlinearity for difference-frequency generation(DFG) into the active region of a dual-wavelength mid-IR QCL using band structure engineering of the QCL active region [Belkin et al., 2009]. However, the disadvantage of DFG QCLs is that they provide less terahertz power than normal QCL. Alternatively, it is also possible to improve temperature performance of QCL by implementation of a second type of waveguide (metal–metal (MM)) design which consists of metal films on both sides of the active region, and provides a mode confinement factor of nearly 100% [Belkin et al., 2009]. Optimized Second Harmonic Generation (SHG) in QC lasers, in which the QWs of the active regions simultaneously function as nonlinear oscillators, have led to high absolute SHG power levels of up to 2 µW and large linear-to-nonlinear power conversion efficiency of around 50–100 µW/W^2[Gmachl et al., 2003].

The present trend in the search of materials for QCL is to use silicon, which can offer the prospect of integrating coherent terahertz (THz) radiation sources with silicon microelectronics. Silicon is the only material that allows mature processing technology which may reduce costs and allow integration with conventional electronic devices. A recent report on silicon based QCL focused on a variety of n-type SiGe-based heterostructures as design candidates [Valavanis et al., 2011]. They have carried out theoretical studies on (001) Ge/GeSi, (111) Si/SiGe, and (001) Si/SiGe material configurations where they have found that (001) Ge/GeSi is the most promising system for development of a Si-based QCL

3.4 Photomixers

Modulated infrared radiation can cause the resonant excitation of plasma oscillations in quantum well diode and transistor structures with high electron mobility. This effect provides a mechanism for the generation of tunable terahertz radiation using photomixing of infrared signals [Preu et al., 2011] .

In terahertz photomixer, the outputs of two continuous-wave sources with frequency difference falling in the terahertz range mix in a photoconductors. The laser induced photocarriers short the gap of device producing photocurrent which is modulated at the laser difference frequency. The generated signal is usually fed to the antenna to transmit into free space. The photomixers are tuneable, coherent, compact, low cost and narrow-band

THz radiation sources. Photomixers have the greatest tuning range among all the coherent sources in terahertz region. Note that photomixing is fundamentally different from difference frequency mixing. Photomixing is more efficient than difference frequency generation at low THz frequencies. At higher THz frequencies, nonlinearities are more efficient due to parasitic impedances which limit the THz bandwidth of photomixers [Mittleman, 2003]. In difference frequency mixing, the outputs of two laser sources are mixed in nonlinear medium like ZnTe or LiNbO₃, where, an intense laser field can cause the polarization of the nonlinear medium to develop new frequency components. These resultant frequency components of polarization act as source of terahertz radiations.

The frequency and power of the generated terahertz signal can be tuned by tuning the central frequency of one of the lasers. Based on model calculations for LT-GaAs photomixer, it has been proposed [Pilla, 2007] that photomixing efficiency can be enhanced by reducing the transit time of majority of carriers in photomixers and photodetectors to < 1 ps. The model indicated that the output terahertz power, P_f is proportional to difference frequency of incident lasers (f) which is given by relation Pf α $f^{-2.77}$ in the 0.5–6.5 THz range.

In difference frequency mixing, an intense pulsed laser induces a significant polarization in non-centrosymmetric structured materials, which can be written as

$$P = \chi^{(1)}E + \chi^{(2)}E \qquad (3)$$

where first term on the right hand side represents linear polarization while second term is responsible for SHG (second harmonic generation), dc rectification, Pockels effect, parametric generation etc

In nonlinear crystals, sum frequency generation (SFG) or difference frequency generation (DFG) can occur, where two pump beams generate another beam with the sum or difference of the optical frequencies of the pump beams. If the incident electric field contains two different frequency components ω_p and ω_s and amplitudes E_g and E_p respectively, the resultant intensity can written as

$$2\chi^{(2)}E_pE_s \{\cos(\omega_p + \omega_s)t + \cos(\omega_p - \omega_s)t\} \qquad (4)$$

It is clearly seen from equation (4) that the second order nonlinear susceptibility will give rise to a nonlinear polarization and reemit radiation at sum ($\omega_p + \omega_s$) and difference ($\omega_p - \omega_s$) frequencies.

Advantages associated with a difference frequency generation include simple frequency control, narrow linewidth and wide tuning ranges based on low loss and phase matched NLO crystals and tunable pump lasers. Diode lasers are sometimes used in difference frequency generation sources for large band width THz signals. Narrow-band terahertz pulse generation via difference frequency mixing using fs laser irradiation of periodically poled lithium niobate (PPLN) and stoichiometric lithium tantalate (PPSLT) crystals[Yu et al., 2011]. The bandwidth of output terahertz radiation obtained was as low as 32 GHz for 1.38 and 0.65 THz for forward and backward THz pulses respectively. The phase mismatch is another critical factor affecting performance of difference frequency generation sources. The conversion efficiency in difference frequency generation process depends primarily on the material's nonlinear coefficient and the phase (velocity) matching conditions. The wave vector of the applied field and the generated field must have the same relation as their frequencies. Unfortunately in nonlinear crystals, the optical beam always travels faster than the THz beam, which makes the phase matching unachievable. Several phase matching techniques been explored to improve output power of terahertz emitters. The phase matching techniques can reduce decay of terahertz radiation in the crystals by techniques like reducing beam diameter [Shibuya et al., 2010] or artificially introducing reversed domain structures (quasi-phase matching, QPM) [Marandi et al., 2008]

4. THz detectors and sensors

4.1 Detectors

THz detectors have been developed based on a variety of principles as described below. However there are two generalized performance parameters that characterize THz detectors. These are Noise Equivalent Power (NEP) and detectivity.

Noise Equivalent Power (NEP) is defined the incident RMS optical power required in order to obtain signal-to-noise ratio of 1 in a bandwidth of 1 Hz. NEP is related to the minimum power that the photodetector can detect

$$NEP = \langle i_n^2 \rangle / R \tag{5}$$

Where i_n is the noise current, and R is the detector responsivity. The related figure of merit is detectivity defined as

$$D^* = \sqrt{SB} / NEP \tag{6}$$

where B is the bandwidth and S is the detector area. [Sizov & Rogalski, 2010]

4.2 Direct and heterodyne THz detection based on the principle of bolometer

Hot electron bolometers are very commonly used for THz detection. The term hot electron describes electrons which are not in thermal equilibrium with the lattice. In metals the heating of electrons does not change the electron mobility and therefore it does not affect the resistance value. The hot electron approach is very productive for semiconductors, where the mobility of electrons depends on their effective temperature. In a superconductor, the hot-electron phenomenon consists of the heating of electrons by radiation. A photon incident on a superconductor is absorbed by a Cooper pair. Due to the large coherence length only one of the electrons of the Cooper pair absorbs the photon. The Cooper pair breaks apart and one highly excited electron with energy close to the incident photon energy and one low-energy quasiparticle are created. Next the quasiparticle loses its energy via electron-electron scattering and creates a secondary excited electron. This process continues so long as the incident radiation is switched on. There are two different techniques to detect the power of submillimeter and far-infrared wavelengths. The first approach is to detect the radiation directly by creating charged carriers or converting the incoming energy to heat proportional to the flux of the incoming photons. The first method is direct detection which is also called an incoherent detection. Direct detectors do not have a fundamental limit of the sensitivity. A direct detector does not produce any noise power unless photons are absorbed. The second method is to shift the incoming radiation to a lower frequency band and then amplify and detect the power of the radiation. The method of shifting the incoming radiation to a lower frequency is called heterodyne principle or coherent detection. In a heterodyne receiver a locally generated frequency is mixed with a signal frequency to produce a signal at a much lower frequency. This frequency conversion is done by a mixer element. The locally generated frequency and the signal frequency are added together to produce a beat frequency, the difference between the signal and the locally generated frequency. The HEB can be used as a heterodyne mixer. The incident radiation and the local oscillator (LO) radiation excite a voltage across the bolometer. The dissipated power of the bolometer is a function of the average absorbed signal power and the average absorbed LO power. The output power consists of a component equal to the difference frequency of the electronic diode. Thus, a HEB can be used as a heterodyne mixer, if it is fast enough to follow the IF [Cherednichenko et al., 2002, Zmuidzinas & Richardds, 2004] .

Superconductor-insulator-superconductor (SIS) tunnel junctions are extremely sensitive heterodyne mixers and have been widely used for frequencies below 1 THz. The upper frequency limit of these devices is determined by the gap frequency of the superconductor defined by fgap = Δ/h with Δ the superconducting gap energy and h the Planck's constant. Drastic losses above the gap frequency of the superconductor (about 700 GHz for niobium) occur inside the superconducting tuning circuit which is used to match the impedance of the SIS device to the antenna impedance. Using superconductors (like NbN or NbTiN) having a higher gap frequency than Nb has increased the upper frequency limit of SISmixers to about 1.2 THz. The main advantage of SIS mixers is the wide Intermediate Frequency (IF) bandwidth provided by this type of mixer. HEB mixers do not have an upper frequency limit and provide a very high sensitivity (T_{rec} < 2000 K) and require very low LO power (< 1 µW). Although the IF bandwidth of HEB mixers is rather limited compared to Schottky and SIS mixers, HEB mixers are the most competitive devices for heterodyne detection in the THz range. When compared by characteristics like response time and dynamic range, SIS detectors are found to be much superior to their TES counterparts [Matsuo, 2006].

The use of low temperatures based on Nb, NbN and a number of related compounds have been reported . In these cases while the fabrication of the bolometer structures is routine since feature sizes are large, the very low operating temperatures (<20 K, in some cases) is a major drawback [Chen et al., 2009; Tretyakov *et al.* ; Stern et al., 2005; Schultz & Lichtenberger, 2007; Ilin et al., 2010; Liu et al., 2009; Kleinschmidt et al., 2007; Cherednichenko et al., 2007] . In addition, these bolometers work in the range < 5 THz, generally. The typical schematic view of a HEB structure is shown in figure 2.

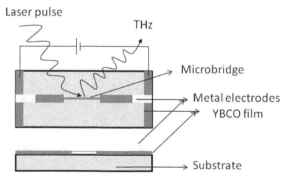

Fig. 2. The typical schematic view of a HEB structure

Bolometers have also shown potential for application in space and astronomy [Meledin et al., 2004; Caumes et al., 2009; Putz et al., 2011]. The discovery of high temperature superconductors has helped in increasing the operating temperature of the HEBs to liquid nitrogen temperatures. For example, an ultrathin PBCO/YBCO/PBCO HEB [Peroz et al., 2007] constructed by ultra-thin (10 nm thick) superconducting micro-bridge coupled with a planar antenna was demonstrated. However, the limitation on mixing operation of HEB is observed causing by aging of thinner layer and nano bridge structure. Tonouchi et al [Tonouchi et al, 2005] have reported extensively on the use of YBCO based antennae for THz detection. There are many other reports on the use of high Tc superconductors as bolometers, but only a few are listed [Du et al, 2011; Aurino et al., 2010; Laviano et al., 2010; Jagtap et al., 2009]

There are some disadvantages of bolometric type of detectors, such as they are extremely sensitive to background radiation, temperature fluctuation, mechanical vibration and electrical interference, and the performance deteriorates with increasing frequencies in the THz range. Further, bolometers are insensitive to phase which does not allow the reconstruction of the pulse shape in the time domain. It should be noted that background radiation defines the limiting condition of sensitivity of detector. The sensitivityof the bolometer is mainly determined by the system noise temperature. System noise includes noise from the input source and noise generated in the receiver.

4.3 Schottky barrier detectors

Schottky diodes have been the only available technology for the detection of THz radiation for a long time {Roeser et al, 1994]. They do not require cooling to cryogenic temperatures and cover a wide frequency range up to several THz. The main disadvantages are the poor sensitivity and the high local oscillator (LO) power requirement. The LO is still a critical element for heterodyne detection in the THz frequency range. Solid state sources often do not provide enough power for the heterodyne operation in the THz range but are under development right now. One of the early reports on Schottky diodes for THz applications was by Suzuki etal [Suzuki et al 1999]. They presented results on submicrometer Pt/GaAs diode fabrication process that resulted in a significant reduction in low-frequency noise. Noise performance was optimized by establishing an even distribution of gallium and arsenic (stoichiometric surface) at the contact surface. Diodes fabricated with the optimized procedure exhibited a signal-to-noise ratio that was better than that of commercially available diodes at 0.9 THz. Rinzan et al [Riznan et al., 2005] demonstrated a heterojunction interfacial work function internal photoemission (HEIWIP) detector with a threshold frequency (f_0) of 2.3 THz (λ_0 = 128 μm). The threshold limit of ~ 3.3 THz (92 µm) was surpassed using AlGaAs emitters and GaAs barriers. The peak values of responsivity, quantum efficiency, and the specific detectivity at 9.6 THz and 4.8 K for a bias field of 2.0 kV/cm are 7.3 A/W, 29%, 5.3×10^{11}, respectively. Yasui et al [Yasui et al, 2006] developed a wide-bandwidth, high-sensitivity, continual terahertz-wave sensor that utilizes a quasioptical parabolic mirror and a Schottky barrier diode and successfully applied it at up to 7 THz range. This sensor utilized a parabolic cylindrical mirror, a long-wire antenna, and a Schottky barrier diode. The antenna, placed at the focal point of the parabolic mirror, quasioptically collects the terahertz signalleading to superior performance. Ryzhii et al., [Ryzhii et al., 2006] investigated the plasma oscillations in a two-dimensional electron channel with a reverse-biased Schottky junction. They show that the plasma instability can be used in a novel diode device - lateral Schottky junction tunneling transit-time terahertz oscillator. Magno et al grew [Magno et al., 2008] Antimonide-based p^+N junctions. The diodes had good rectification with ideality factors near 1, and high saturation current densities of 2.5×10^{-2} A/cm^2. A cutoff frequency of 6.5 THz is estimated from the RC product. The high saturation current indicates that this diode could be used as a terahertz mixer. Maestrini et al [Maestrini et al, 2010] reviewed some aspects of the technology of terahertz heterodyne receiver front-ends dedicated to astrophysics, planetary and atmospheric sciences with focus on frequency multipliers and Schottky mixers. It is clear from this review that there is much potential for research in the area of novel architectures of power-combined frequency multipliers at submillimeter-wavelengths, THz planar fundamental mixers, and integrated receivers and the fabrication of submillimeter-wave planar Schottky diodes. Chen et al. [Chen et al, 2010] demonstrated high performance nanometer NiSi$_2$-Si Schottky barrier diode arrays (SBDA) with various isolation designs, including poly Si gate (PSG) and resist protection oxide (RPO) for advanced radio frequency applications. The SBDAs could achieve a cutoff frequency of up to 4.6 THz.

It is evident from a brief survey of literature that Schottky diode structures have great potential for application in THz detectors impacting a variety of fields. However, there are several challenges in design, fabrication and operation of these device structures that remain unresolved

4.4 Field-effect transistor detectors

A channel of a Field Effect Transistor (FET) can act as a resonator for plasma waves. The plasma frequency of this resonator depends on its dimensions and for gate lengths of a micron and sub-micron (nanometer) size, can reach the Terahertz (THz) range {Dyakanov, 2008]. The interest in the applications of FETs for THz spectroscopy started at the beginning of 90s with the work of Dyakonov and Shur [Dyakanov and Shur, 1996] who predicted that plasma oscillations in a FET channel can lead to the THz emission. Recently, non-resonant plasma properties were successfully used for the room temperature broadband THz detection and imaging by silicon FETs. Both THz emission and detection, resonant and non-resonant, were observed experimentally at cryogenic, as well as at room temperatures, clearly demonstrating effects related to the excitation of plasma waves. The possibility of the detection is due to nonlinear properties of the transistor, which lead to the rectification of an ac current induced by the incoming radiation. As a result, a photoresponse appears in the form of dc voltage between source and drain which is proportional to the radiation power (photovoltaic effect). There are three distinct regions of operation of the FETs depending on gate lengths. It has been demonstrated that for gate lengths of the order of 0.1 µm the plasma oscillations will be in the low THZ region [Dyakanov, 2008]. Otherwise the FET can operate as a broad-band THz detector. There are a few other reports on the use of FETs as THz detectors, signifying again the potential of these structures as wellas the number of fabrication issues that need to be overcome [Knap et al., 2011]. The use of nitride based FETs for THz detection has also been demonstrated.[Shur, 2007]

4.5 High electron mobility transistor detectors

Semiconductors heterostructure devices such as high electron mobility transistor (HEMT) and heterojunction bipolar transistors have attracted increasing attention in THz devices

[Dyakonov 2005; Fatimy 2005; Knap 2004; Lusakowski 2005; Otsuji et al., 2010; Tonouchi 2007; Tredicucci 2005], which in fact, can be regarded as a new type of terahertz detectors. Otsuji *et al.* [Otsuji et al., 2010] reviewed recent advances in emission of terahertz radiation from 2D electron systems in semiconductor heterostructures. A self-consistent spatiotemporal variation of quantum confined electron gas with sufficiently high electron mobility in the channel of HEMT has been proved to be an efficient mechanism for the detection and emission of terahertz radiations. The main advantage of the HEMT over the non-laser based sources is in working temperature and hence hardware requirement for operation. Another major advantage of HEMT terahertz emitters are their high tunability and low fabrication cost. The introduction of closely-spaced heterojunction interfaces under the right conditions produces a two-dimensional electron gas (2 DEG). This low dimensional plasmon (electron gas) is favourable for solid-state far-infrared and THz sources. The strong confinement of charge carriers in the quantum well is believed to be the cause of large electron mobility. Further, the small channel length of the device, leading to an increase in transconductance with an increase in carrier mobility, is well known for HEMT. However, the high losses caused by free carrier absorption and small confinement factor of heterostructure are the factors which restricts use of HEMT at longer THz wavelengths. Monte Carlo simulation performed to study of transient electron transport in wurtzite GaN, InN, and AlN shown that, for ideal transistors with L = 0.1 µm, by replacing the GaN channel with InN, can be expected increase frequency of operation from 0.48 to 0.68 THz. Moreover, low m_e (effective electron mass) in the InN channels also gives a prospect for electron nonstationary dynamics, and operating frequency over 1 THz has been predicted for 0.1 µm gates in that case [Foutz et al., 1999]. Further increase in operating frequency (THz range) is, in fact, expected in ultrathin gate structured HEMT. Fatimy *et al.* [Fatimy et al., 2010] have demonstrated room temperature terahertz generation by a submicron size AlGaN/GaN-based high electron mobility transistor. They have showed that these transistors are tunable by the gate voltage between 0.75 and 2.1 THz. It is also possible to tune operating frequency by manipulating cap, window, and gate lengths and other structural parameters [Ryzhii et al., 2006].

Understanding the physics of plasma wave resonant line broadening is essential to improve performance of HEMT based plasma wave detectors of terahertz radiation. Properties of HEMT terahertz detectors are defined by plasma wave excitations of two dimensional electron gas [Tsaur & Wang, 2009]. Recently it has been demonstrated that narrow terahertz

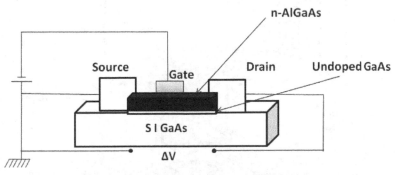

Fig. 3. The typical schematic view of a high electron mobility transistor structure

plasma wave resonant detection at low temperature in 200 nm gate length InGaAs/InAlAs multichannel HEMT. To achieve selective resonant and voltage tunable terahertz emission, they have maintained the gate width is in the order of the gate length [Shchepetov et al., 2008].The typical schematic of a High electron mobility transistor used for THz detection is shown in fig.3.

4.6 Sensors

The state of technology in THz sensors is briefly reviewed. A variety of THz sensors have been developed recently mainly for use in biological and medical applications . Federici et al have reviewed the developments in THz sensing [Federici et al, 2005] technology, spectroscopy and imaging for security applications. The typical view of a THz imaging set up is shown in fig.4. The authors state that since, (a) THz radiation can detect concealed weapons since many non-metallic, non-polar materials are transparent to THz radiation; (b) target compounds such as explosives and illicit drugs have characteristic THz spectra that can be used to identify these compounds and (c) THz radiation poses no health risk for scanning of people stand-off interferometric imaging and sensing for the detection of explosives, weapons and drugs is critical. Future prospects of THz technology are discussed. Nagel et al [Nagel et al, 2003] first reported a novel resonant THz sensor for the label-free analysis of DNA molecules. The sensor allows the direct detection of DNA-probe molecules at functionalized electrodes via hybridization. Subsequent time resolved photoconductive sampling of the THz transmission identifies the binding state between probe and target DNA. Integrating neighbouring sensors on a chip, this technique can be extended to a parallel analysis of multiple DNA sequences. Nagel et al [Nagel et al, 2006] then reviewed the state of biological THz technology applications and conclude that success in this area will strongly depend on the development of compact, low-cost and flexible systems. They have reported on different approaches for THz biosensor systems based on femtosecond lasers and discussed the technology for generation, transmission and detection of THz signals as well as their application formarker-free biomolecule detection on functionalized

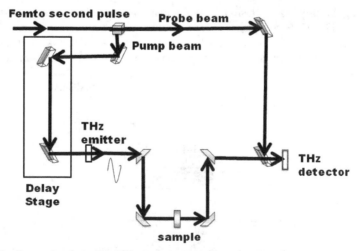

Fig. 4. Typical schematic view of a THz imaging set up.

surfaces in dry and fluid environments. Kitagawa et al [Kitagawa et al , 2006] demonstrated the THz spectroscopic performance of highly integrated sensor chips based on microstrip lines by measuring biomolecule/water systems. The concentration resolution of the chips reaches down to 0.05 g/ml. They have confirmed that the number of bound water molecules per biomolecule can be obtained with precision using solid state transmission lines. The chips are highly suitable for the inspection of small amounts of specimen and for the application to a wide range of water rich materials. McLaughlin et al [McLaughlin et al , 2008] employed poled polymer films 10–15 μm thick with electro-optic coefficients as high as 160 pm/V at 1300 nm are used to generate and sense subpicosecond pulses with continuous bandwidth up to 15 THz. The use of the poled polymer as the terahertz sensor for the identification of DAST phonons at 1.1, 3.0, 5.3, 8.5, and 12.5 THz was demonstrated. Kawase etal [Kawase etal , 2010] have suggested a wide range of real-life applications using novel terahertz imaging techniques. A high-resolution terahertz tomography has been demonstrated by ultra short terahertz pulses using optical fiber and a nonlinear organic crystal. They, further, describe a nondestructive inspection system that can monitor the soot distribution in a ceramic filter using millimeter-to-terahertz wave computed tomography. Another proposed application is in the area of thickness measurement of very thin films using the high-sensitivity metal mesh filter. Abbas et al [Abbas et al , 2009] describe the development, functionalization and functionality testing of a TeraHertz (THz) Bio-MicroElectroMechanical System (BioMEMS) dedicated to enzyme reaction analysis. The microdevice was fabricated by mixing clean room microfabrication with cold plasma deposition. The progression of the hydrolysis reaction over time was monitored by the THz sensor connected to a vectorial network analyzer. Preliminary results showed that sub-THz transmission measurements are able to discriminate different solid films, various aqueous media and exhibit specific transmission behavior for the enzyme hydrolysis reaction in the spectral range 0.06–0.11 THz. Rau et al [Rau et al., 2005] reported on a simple THz waveguide element used as a sensitive sensor for adsorbates on surfaces. The evanescent wave from total internal reflection off a silicon prism was used to couple pulsed THz radiation frequency selectively into the waveguide. The coupled frequencies were determined via time domain spectroscopy and react sensitively to any changes of thickness or phase shift upon reflection. In particular, the sensitivity to phase shifts makes this waveguide sensor attractive for the detection of very thin adsorbates. Xiao-li and Jiu-sheng [Xiao-li & Jiu-sheng,2011] used terahertz time-domain spectroscopy technique to test flour/talc powder mixture samples. They found that the two samples have distinct absorption peaks and different refractive index in the THz region leading to the possibility of THz applications to the field of food safety. Mendis et al [Mendis et al, 2009] describe a terahertz optical resonator suited for highly sensitive and noninvasive refractive-index monitoring. The resonator is formed by machining a rectangular groove into one plate of a parallel-plate waveguide, and is excited using the lowest-order transverse-electric (TE$_1$) waveguide mode. Since the resonator can act as a channel for fluid flow, it can be easily integrated into a microfluidics platform for real-time monitoring. Using this resonator with only a few microliters of liquid, a refractive-index sensitivity of 3.7×10^5nm/refractive-index-unit was demonstrated in the THz range. Lu et al [Lu et al., 2008] discuss broadband terahertz wave detection through field-induced second harmonic generation using selected gases. The dependences of the detected second harmonic intensity on probe pulse energy, bias field strength, gas pressure, and third order nonlinear susceptibility are systematically investigated with xenon, nitrogen, SF$_6$, and alkanes. Experiment results reveal that the detected second harmonic intensity quadratically depends on the third order nonlinear susceptibility of the gas. Two orders of magnitude enhancement in the dynamic range of broadband terahertz wave detection are observed with alkane gas (C_4H_{10}) sensor. Foltynowicz et al [Foltynowicz et al, 2006] reported the vapor-phase spectrum of 2,4-dinitrotoluene (DNT) from 0.05 THz to 2.7 THz utilizing pulsed terahertz time-domain spectroscopy. They observed a broad-band absorption profile from 50 GHz to 600 GHz, peaking at 240 GHz. This broad absorption profile corresponds to DNTs pure rotational spectrum, which was confirmed by asymmetric top model calculations. Superimposed on the broad absorption profile, we observed discrete structure that extended to the higher THz frequencies. Leahy-Hoppa et al [Leahy-Hoppa et al, 2007] have shown Time-domain terahertz spectroscopy (TDTS) to be a promising tool in detection of explosives and explosive related compounds. The THz absorption spectra over an extended frequency band from 0.5 to 6 THz were reported for four explosives: RDX (1,3,5-trinitroperhydro-1,3,5-triazine), HMX (1,3,5,7-tetranitroperhydro-1,3,5,7-tetrazocine), PETN (pentaerythritol tetranitrate), and TNT (2,4,6-trinitrotoluene). New distinctive spectral features are shown in these materials between 3 and 6 THz. Zhang et al [Zhang et al, 2008] present terahertz reference-free phase imaging for identification of three explosive materials (HMX, RDX, and DNT). They propose a feature extraction technique to locate the spectral position of an unknown material's absorption lines without using the reference signal. The samples are identified by their absorption peaks extracted from the negative first-order derivative of the sample signal phase divided by the frequency at each pixel. Hu etal [Hu et al, 2006] report experimental measurement and theoretical analysis of THz spectrum for five different explosives and related compounds are introduced. The refractive index and absorption coefficient of the samples are measured in the region of 0.2–2.6 terahertz by time-domain spectroscopy. The simulated spectrum of γ-HNIW is in agreement with the experimental data.

It is thus evident that there are many significant developments in the area of THz sensors that impact various aspects of biology, medicine and security.

4.7 New materials and device structures

A large variety of new types of materials and device structures have been proposed recently for THz applications. These include the use of carbon nanotubes [Costa et al., 2009], GaAs based nanotransistors [Łusakowski, 2007] and polar optical phonons in quantum wells [Liu et al., 2011].Sputter deposited zinc oxide photoconductive antenna have also been proposed for THz detection applications[Iwami et al., 2009]. Videlier et al have demonstrated Si MOSFET structures as THz detectors[Videlier et al].The possibility of utilizing two-dimensional plasmons in semiconductor heterostructures for THz detection has been also been proposed [Otsuji et al, 2010]. Cox et al [Cox et al., 2011} demonstrated a MEMS based uncooled THz detector. The development of novel single photon detectors that work in the range of 10-50um based on double quantum weel structures have been reported recently {Ueda& Komiyama, 2010]. There are many other reports on novel materials and structures, these are however beyond the scope of the current review.

5. ZnTe coatings for THz applications

In this section we present a review of the work being carried out by the present authors in the area of THz materials development. The ability to miniaturize THz photonics depends critically on two developments compact THz sources and elimination of the dependence on crystals as emitters and detectors. The focus is, therefore, mainly on the development of thick ZnTe coatings for use in THz emitter and detector applications, based on the principle of optical rectification. As stated in section 3.2, optical rectification can be controlled by manipulating the crystal structure and orientation. It is, however, a weak non-linear response. Hence, material thicknesses have to be high to generate a measurable response. Semiconductors from the II-VI group, especially ZnTe crystals [Turchinovich & Dijkhuis, 2007;], have drawn considerable interest in recent years for THz applications. ZnTe is intrinsically a p-type semiconductor from the II-VI family with a direct bandgap of 2.26 eV. The growth of *n*-type ZnTe is still a challenge, which must be overcome before *p-n* junction devices can be realized. Incorporation of excess Zn during growth may increase donor concentration favoring the formation of *n*-type ZnTe. The more commonly encountered polymorph of ZnTe is the zinc blende structure which has four asymmetric units in its unit cell as against two in the wurtzite form. The higher degree of asymmetric coordination in zinc blende ZnTe favors formation of non-centrosymmetric structure which is very useful in terahertz generation and detection applications [4]. The applications of wurtzite ZnTe, on the other hand, are less well reported. Information on properties like spectral transmission, refractive index, optical band gap, and electronic structure of wurtzite ZnTe thin films is very sparse.

Since most of the THz related work on ZnTe is on single crystals, the main objectives of our work is to fabricate ZnTe coatings that are 10-12 microns thick as well as stabilizing the wurtzite form of ZnTe and investigate its optical and electronic properties. The coatings were produced by either thermal or electron beam evaporation in high vacuum (10⁻⁶ Torr). The source material is zinc blende or wurtzite ZnTe. The substrates are borosilicate glass or fused silica and the films were deposited at room temperature. Spectral transmittance curves of the films in the wavelength range from 190 to 2500 nm were measured using a spectrophotometer and from these spectra the refractive index and optical band gaps were extracted. Crystal structure was determined suing a powder x-ray diffractometer and electron diffraction while microstructure and morphology was examined under an Atomic force microscope.

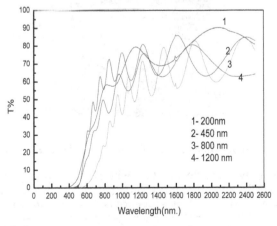

Fig. 5. Spectral transmission curves

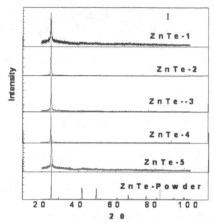

Fig. 6. X-ray diffraction patterns of the ZnTe films of different thicknesses showing the formation of zinc blende structure.

The measured spectral transmission curves for zinc blende ZnTe films of thicknesses upto 1200 nm and typical x-ray diffraction pattern of the 1200 nm film is shown in figs. 5 and 6 respectively. The transmission of the films is close to 90% in the near IR region and calculated refractive indices were between 2.0 and 2.6. The optical band gaps were between 2 and 2.2 eV. These values compared favourably with earlier reports.Significantly the 1200 nm thick film was nanocrystalline (with a crystallite size of 24 nm) and showed a preferred (111) orientation. The thickest films achieved using this technique had a thickness of 10-12 microns

The atomic force microscope image of a 2000 nm thick zinc blende ZnTe film is shown in fig. 7. The film is clearly very smooth with roughness less than 10 nm even at this thickness. The grain sizes are of the order of 100 nm.

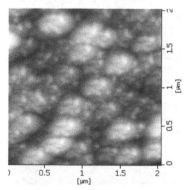

Fig. 7. Atomic force microscope image of a 2000 nm zinc blende ZnTe film.

ZnTe also crystallizes in the wurtzite structure wherein the stacking sequence of the cation and anion is ABABABAB. . . with each repeated period perpendicular to the basal plane, whereas in the zinc blende structure the sequence is ABCABCABC. Therefore, the distance between the third nearest-neighbor (NN) atoms is small in wurtzite ZnTe. In contrast, the distance between the first and second NN atoms is invariable in both structures. In addition, the Zn-Te bond in zinc blende ZnTe has greater ionic character than that in wurtzite ZnTe. Due to these differences it has been shown from ab-initio calculations that the band gaps of the structures should be different. Similarly, it is also expected that the electronic properties will be different. There are, however, very limited studies on the growth and stabilization of wurtzite ZnTe films. The present authors have grown wurtzite ZnTe films by electron beam evaporation of a wurtzite ZnTe source material. Typical x-ray diffraction patterns of the wurtzite films are shown in fig.8 for different thicknesses. The measured spectral transmittance for the zinc blende and wurtzite films are shown in fig.9. Detailed dispersion analysis of the films and comparison with zinc blende films has been carried out. ZnTe thin films exhibit either the wurtzite or zinc blende structure depending on the chemical nature of the source material. The formation of the structural phases is confirmed by a combination of x-ray diffraction and selected area electron diffraction. The films were nanocrystalline and randomly textured. The refractive index of the wurtzite phase (3.87 at 2000 nm) is higher than that for the zinc blende phase (3.03 at 2000 nm). The optical band gap, however, is lower for the wurtzite phase (1.1 eV) than the zinc blende phase (2.26 eV). The optical dispersion behavior has been investigated within the framework of the single effective oscillator model. It is shown that the oscillator strength, dispersion energy and optical charge carrier concentrations are all lower for the wurtzite phase than the zinc blende phase.

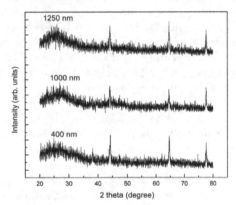

Fig. 8. XRD patterns of wurtzite ZnTe films of different thickness

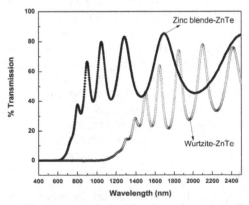

Fig. 9. Comparison of optical response of zinc blende and wurtzite ZnTe thin films.

The charge carrier concentration in wurtzite phase ($5 \times 10^{25} m^{-3}$) is lower than the zinc blende structure (4.17×10^{26} m^{-3}) This may make the wurtzite phase suitable for the development of n-type ZnTe, which has been a long standing challenge. Evidently the wurtzite phase has the potential to be employed as an n-type material leading to the formation of ZnTe p-n junctions for THz applications. Further developments in this direction are ongoing

Our work thus demonstrates the growth of thick ZnTe coatings in both the zinc blende and wurtzite polymorphs. Their optical properties are substantially different, signifying the possibility of applications in different frequencies regimes of the THz region [Kiran et al, 2010; Kshirsagar et al, 2011].

6. Conclusions and future prospects

In summary, a number of technologies for THz emitters, detectors and sensors have been reviewed. It is obvious that, while much progress has been made in all three areas, there is scope for new science and technology to emerge in all these areas. Significantly, the progress in the area of THz electronics has been much more rapid than that in the area of THz photonics. This is mainly due to the lack of cost effective so-called „table top" femto second laser sources. THz electronics, on the other hand, relies solely on existing mature technologies for materials and device fabrication. Much of the materials development has focused on single crystals. However, for miniaturization of THz technologies (Photonics and electronics) to succeed, the area of coatings, films and micro-nano devices has to be focused upon, in addition to development of compact THz sources. A report on the development of a high-resolution terahertz (THz) imaging beyond the diffraction limit by using a two-dimensional electron gas (2DEG) in a GaAs/AlGaAs heterostructure, points to effort in this direction[.Kawano & Ishibashi, 2010]. In the area of materials, strongly correlated electron systems have shown great potential[Ge et al, 2010]. In addition, novel device structures to exploit the plasma instability for THz emission and detection is another area that is receiving greater attention. Interestingly, access to fs laser sources has led to better understanding of physical and chemical phenomena in the THz region, that will eventually lead to new materials and device structures. Two examples exemplify this theme. It has recently been shown [Chan et al., 2011] that a nano mechanical oscillator coupled to single mode electromagnetic oscillator in a nano optical cavity can be designed to have oscillations in THz frequency domain. The physics of this domain of energy storage as envisaged by the Debye model has to be examined and the modes of vibrations at phase boundaries, and domain boundaries studied through THz experiments. These will reveal new materials and methods of generation and detection and sensing in the THz range. In another report [Chen et al, 2011], impulsive stimulated Raman scattering and Time-domain THz spectroscopy of coherent optical phonons in bismuth germinate ($Bi_4Ge_3O_{12}$) revealed more than 12 unique vibrational states ranging in frequency from 2 to 11 THz, each with coherent lifetimes ranging from 1 to 20 ps. Such studies on other systems can be expected have a profound impact on the developments in THz sources, emitters, detectors and sensors.

7. Acknowledgements

The authors acknowledge the facilities provided by the School of Physics under the UGC-UPE and CAS programs. A Dr D S Kothari Fellowship for SK is also acknowledged.

8. References

Abbas, A.; Treizebre, A.; Supiot, P.; Bourzgui, N.; Guillochon, D.; Vercaigne-Marko, D. & Bocquet, B. (2009). Cold plasma functionalized TeraHertz BioMEMS for enzyme reaction analysis, *Biosensors and Bioelectronics*, Vol.25, No.1, (September 2009), pp. 154-160, ISSN 09565663

Aurino, M., Kreisler, A.J., Türer, I., Martinez, A., Gensbittel, A., Dégardin, A.F. (2010) *Journal of Physics: Conference Series* Vol.234, No.4 (2010) art. no. 042002. ISSN: 1742-6588.

Belkin, M. A.; Jie, W. Q.; Pflugl, C.; Belyanin, A.; Khanna, S. P. Davies, A. G.; Linfield, E. H. Capasso, F. (2009). High-Temperature Operation of Terahertz Quantum Cascade Laser Sources, *IEEE Selected Topics in Quantum Electronics*, Vol.15, No.3 (June 2009) pp. 952-967, ISSN 1077-260X

Bhasin L. & Tripathi, V. K. (2009). Terahertz generation via optical rectification of x-mode laser in a rippled density magnetized plasma, *Phys. Plasmas* Vol.16, No. 10, (October 2009), pp. 103105- 1-6 ISSN 1089-7674

Blanchard,F.; Sharma, G.; Razzari, L.; Ropagnol, X.; Bandulet, H –C.; Vidal, F.; Morandotti, R.; Kieffer, J –C.; Ozaki, T.; Tiedje, H.; Haugen, H.; Reid, M. & Hegmann, F. (2011). Generation of Intense Terahertz Radiation via Optical Methods, IEEE Selected Topics in Quantum Electronics, Vol.17 No. 1, (June 2011), pp. 5-16, ISSN 1077-260X

Bolivar P H, Nagel M, Richter F, Brucherseifer M, Kurz H, Bosserhoff A & Buttner R (2004), Label-free THz sensing of genetic sequences: towards 'THz biochips', *Philosophical Transactions of the Royal Society, London , A., Vol.* 362 (2004) pp.323-335.

Bugay, A. N. & Sazonov, S. V. (2010). The generation of terahertz radiation via optical rectification in the self-induced transparency regime, *Phys. Lett. A*, Vol.374, No. 8 (February 2010), pp. 1093-1096, ISSN 0375-9601

Caumes, J. P.; Chassagne, B.; Coquillat, D.; Teppe, F. & Knap, W. (2009). Focal-plane micro-bolometer arrays for 0.5 THz spatial room-temperature imaging, *IEEE Electronics Letters*, Vol.45, No. 1, (December 2009), pp. 34-35, ISSN 0013-5194

Chan J., Mayer Alegre T.P., Safavi-Naeini A.H.,, Hill J.T., Krause A, Groblacher S., Aspelmeyer M & Painter O, (2011), Laser cooling of a nanomechanical oscillator into its quantum ground state, *Nature*, Vol. 478, (October, 2011), pp. 89-92. ISSN: 0028-0836.

Chamberlain J.M., (2004) Where optics meets electronics: recent progress in decreasing the terahertz gap, *Philosophical Transactions of the Royal Society, London , A., Vol.* 362 (2004) pp.199–213. ISSN: 1471-2962.

Chen, J.; Liang, M.; Kang, L.; Jin, B. B.; Xu, W. W.; Wu, P. H.; Zhang, W.; Jiang, L.; Li, N. S. & Shi, C. (2009). Low Noise Receivers at 1.6 THz and 2.5 THz Based on Niobium Nitride Hot Electron Bolometer Mixers, *Appl. Superconductivity, IEEE Trans.* Vol.19, No. 3, (July 2009), pp. 278-271, ISSN 1051-8223

Chen, S.-M.; Fang, Y.-K.; Juang, F.R.; Yeh, W.-K.; Chao, C.-P. & Tseng, H.-C (2010). Terahertz Schottky barrier diodes with various isolation designs for advanced radio frequency applications, *Thin Solid Films*, Vol.519, No.1, (October 2010), pp. 471-474, ISSN 0040-6090.

Chen, Z.; Gao, Y.; & DeCamp, M. F. (2011). Retrieval of terahertz spectra through ultrafast electro-optic modulation, *Appl. Phys. Lett.*, Vol.99, No.1, (July 2011), pp. 011106-09 ISSN 1077-3118

Cherednichenko S., Khosropanah P., Kollberg E., Krong M., Merkel H., (2002), Terahertz superconducting hot electron bolometer mixers, *Physica C; Superconductivity*, Vol. 372-376, pt. 1, (August 2002), pp. 407-415.

Coquillat, D., Teppe, F., Videlier, H., Coquillat, D., Lusakowski, J., Skotnicki, T. Silicon field effect transistors for Terahertz detection and imaging Knap, W., Shuster, F., (2011) *Proceedings of the 5th European Conference on Antennas and Propagation, EUCAP 2011* , art. no. 5782258, (2011) pp. 3180-3182.

Costa M. R., Kibis O.V., & Portnoi M.E.(2009), Carbon nanotubes as a basis for terahertz emitters and detectors, *Microelectronics Journal* , Vol. 40 (2009) 776–778

Cox, J.A., Higashi, R., Nusseibeh, F., Zins, C. (2011) MEMS-based uncooled THz detectors for staring imagers *Proceedings of SPIE - The International Society for Optical Engineering* 8031, (2011) art. no. 80310D.

Du, J., Hellicar, A.D., Hanham, S.M., Li, L., MacFarlane, J.C., Leslie, K.E., Foley, C.P. (2011) YBCO hot-electron bolometers dedicated to THz detection and imaging: Embedding issues *Journal of Infrared, Millimeter, and Terahertz Waves* Vol. 32 No.5, (2011) pp. 681-690.

Dyakonov M. I,.(2010) Generation and detection of Terahertz radiation by field effect transistors *Comptes Rendus Physique*, Vol. 11 (2010) pp. 413–420

Dyakonov, M.; & Shur, M. (1996). Detection, mixing, and frequency multiplication of terahertz radiation by two-dimensional electronic fluid, (August 1996), *Electron Devices, IEEE Trans*, Vol.43, No.3, (1996) pp. 380-387, ISSN 0018-9383

Dyakonova, N.; Teppe, F.; Lusakowski, J.; Knap, W.; Levinshtein, M.; Dmitriev, A.P.; Shur, M.S.; Bollaert, S.; Cappy A. (2005). Magnetic field effect on the terahertz emission from nanometer InGaAs/AlInAs high electron mobility transistors, *J. Appl. Phys.* Vol.97, No.11, (May 2005), pp. 114313-1-3 ISSN 1089-7550

Fatimy, A. E.; Dyakonova, N.;, Meziani, Y.; Otsuji, T.; Knap, W.; Vandenbrouk, S.; Madjour, K.; Théron, D.; Gaquiere, C.; Poisson, M. A.; Delage, S.; Prystawko, P. & Skierbiszewski C.; (2010). AlGaN/GaN high electron mobility transistors as a voltage-tunable room temperature terahertz sources, *J. Appl. Phys.* Vol.107, No.2, (January 2010), pp-024504 1-4, ISSN 1089-7550

Federici, J. F.; Schulkin, B.; Huang, F.; Gary, D.; Barat, R.; Oliveira, F. & Zimdars, D. (2005). THz imaging and sensing for security applications—explosives, weapons and drugs, *Semicond. Sci. Technol.* Vol.20, No.7 (July 2005), pp. S266, ISSN 0268-1242

Ferguson B., Zhang X.C.,(2002) Materials for Terahertz Science and Technology, *Nature Materials* Vol. 1 (2002) 26-33. ISSN 1476-1122.

Foltynowicz, R. J.; Allman, R. E. & Zuckerman, E. (2006). Terahertz absorption measurement for gas-phase 2,4-dinitrotoluene from 0.05 THz to 2.7 THz, *Chemical Physics Letters*, Vol.431, No.1 (November 2006), pp 34-38, ISSN 0009-2614

Foutz, B. E.; O'Leary, S. K.; Shur, M. S. & Eastman, L. F. (1999). Transient electron transport in wurtzite GaN, InN, and AlN, *J. Appl. Phys.*, Vol.85, No.11 (February 1999), pp. 7727 1-8, ISSN 1089-7550

Freeman, J. R.; Brewer, A.; Beere, H. E. and Ritchie, D. A., (2011). Photo-luminescence study of heterogeneous terahertz quantum cascade lasers, *J. Appl. Phys.*, Vol.110, No.1, (July 2011) pp. 013103 1-6, ISSN 1089-7550

Ge C., Jin K., Lu H., Wang C., Zhao G., Zhang L., Yang G., (2010), Mechanisms for the enhancement of the lateral photovoltage in perovskite heterostructures, *Solid State Communications*, Vol. 150 (2010) pp. 2114–2117

Ginzburg N. S., Malkin A.M., Yu. Peskov N., Sergeev A. S., Yu. Zaslavsky V,& Zotova I.V.,(2011) Powerful terahertz free electron lasers with hybrid Bragg reflectors, *Physical Review Special Topics: Accelerators and Beams* Vol.14, No.4 (2011) 042001-1-9.

Gmachl, C.; Belyanin, A.; Sivco, D.L.; Peabody, M.L.; Owschimikow, N.; Sergent, A.M.; Capasso, F. & Cho, A.Y. (2003). Optimized second-harmonic generation in quantum cascade lasers, *Quantum Electronics, IEEE,* Vol.39 , No.11, (November 2003), pp. 1345-1355, ISSN 0018-9197

He, S.; Chen, X.; Wu, X.; Wang, G. & Zhao, F. (2008). Enhanced Terahertz Emission From ZnSe Nano-Grain Surface, *Journal of Lightwave Technology,* Vol.26, No.11, (June 2008), pp. 1519 – 1523, ISSN 0733-8724

Hoffmann, M. C.; Yeh, K.; Hwang, H. Y.; Sosnowski, T. S.; Prall, B. S.; Hebling, J. & Nelson, K. A. (2008). Fiber laser pumped high average power single-cycle terahertz pulse source, *Appl. Phys. Lett.* Vol.93, No. 14 (October 2008), pp. 141107 1-3 ISSN 1077-3118

Hu, Y.; Huang, P.; Guo, L.; Wang, X. & Zhang, C. (2006). Terahertz spectroscopic investigations of explosives, *Phys. Lett. A,* Vol.359, No.6, (December 2006), pp. 728-732, ISSN 0375-9601

Il'in, K.S.; Semenov, A.D.; Hübers, H.-W. & Siegel, M. (2010). Hot-electron bolometer mixers for terahertz radiation, *Electron. Lett.* Vol.46, No. 26, (January 2010), pp. S14-S16, ISSN 0013-5194

Innocenzi P, Malfatti L, Piccinini M, Sali D, Schade U,& Marcelli A (2009), Application of Terahertz Spectroscopy to Time-Dependent Chemical-Physical Phenomena, *Journal of Physical Chemistry A,* Vol. 113, No. 34, (2009) pp. 9418–9423.

Iwami, K., Ono, T., Esashi, M. 2009,Sputter deposited zinc oxide photoconductive antenna for terahertz time-domain spectroscopy *Proceedings of SPIE - The International Society for Optical Engineering* Vol. 7133, (2009), art. no. 713315

Jacobsen R. H., Mittleman D. M.,& M. C. Nuss, (1996) Chemical recognition of gases and gas mixtures with terahertz waves, *Optics Letters,* Vol. 21, No. 24 (December 1996) pp.2011-2013.

Jagtap V.S., Scheuring A., Longhin M., Kreisler A.J., Dégardin A.F., (2009) From superconducting to semiconducting YBCO thin film bolometers: Sensitivity and crosstalk investigations for future thz imagers *IEEE Transactions on Applied Superconductivity* Vol. 19, No. 3 (2009) pp. 287-292.

Kasai, S.; Katagiri, T.; Takayanagi, J.; Kawase, K. and Ouchi, T. (2009). Reduction of phonon resonant terahertz wave absorption in photoconductive switches using epitaxial layer transfer, *Appl. Phys. Lett.,* Vol.94, No.11, (2009) pp. 113505 1-3 ISSN 1077-3118

Kawano Y., Ishibashi K. (2010) Physica E 42 (2010) 1188–1191

Kawase, K.; Shibuya, T.; Hayashi, S. & Suizu, K. (2010). THz imaging techniques for nondestructive inspections, *Comptes Rendus Physique,* Vol.11, No.7, (October 2010), pp. 510-518, ISSN 1631-0705

Kiran M.S.R.N., Kshirsagar S.D., Krishna M.G., Tewari S.P. (2010), Structural, optical and nanomechanical properties of (111) oriented nanocrystalline ZnTe thin films, European Journal of Applied Physics: Applied Physics, vol. 51 (October 2010) pp.10502-9.

Kitagawa, J.; Ohkubo, T.; Onuma, M. & Kadoya, Y., (2006). THz spectroscopic characterization of biomolecule/water systems by compact sensor chips, *Appl. Phys. Lett.* Vol.89, No.4, (July 2006), pp. 041114 1-3, ISSN 1077-3118

Klatt G., Hilser F., Qiao W., Beck M., Gebs R., Bartels A., Huska K., Lemmer U., Bastian G., Johnston M.B., Fischer M., Faist J., & Dekorsy T., (2010) Terahertz emission from lateral photo-Dember currents, Optics Express, Vol. 18, Issue 5, (2010)   pp. 4939-4947

Kleinschmidt, P.; Giblin, S. P.; Antonov, V.; Hashiba, H.; Kulik, L.; Tzalenchuk, A. & Komiyama, S. (2007). A Highly Sensitive Detector for Radiation in the Terahertz Region, *IEEE trans. Instru. measurement,* Vol.56, No. 2, (2007) pp. 463-467, ISSN 0018-9456

Kida N, Murakami H & Tonouchi M., (2005) Terahertz optics in strongly correlated electron systems, *Topics in Applied Physics,* Vol. 97 (2005) pp. 215-334.

Knap, W.; Lusakowski, J.; Parenty, T.; Bollaert, S.; Cappy, A.; Popov, V.V. & Shur, M.S. (2004). Terahertz emission by plasma waves in 60 nm gate high electron mobility transistors, *Appl. Phys. Lett.* Vol.84, No.13, (February 2004), pp. 2331 1-3, ISSN 1077-3118

Knap, W.; Shuster, F.; Coquillat, D.; Teppe, F.; Videlier, H.; Coquillat, D.; Lusakowski, J. & Skotnicki, T. (2011). Silicon field effect transistors for Terahertz detection and imaging, *Proceedings of the 5th European Conference on Antennas and Propagation, Antennas and propagation. european conference,* April, 2011, Rome, Italy

Krotkus A. (2010), Semiconductors for terahertz photonics applications, *Journal of Physics D ; Applied Physics, Vol.43* (2010) pp.273001-273006. ISSN : 0022-3727

Kshirsagar S D, Krishna M.G., Tewari S.P., (2011), Morphological and Optical properties of Wurtzite ZnTe thin films, *AIP Conference proceedings,* Vol. 1349 (July 2011), pp. 1285-6.

Laviano, F., Gerbaldo, R., Ghigo, G., Gozzelino, L., Minetti, B., Rovelli, A., Mezzetti, E., (2010) Rugged superconducting detector for monitoring infrared energy sources in harsh environments, *Superconductor Science and Technology* Vol. 23, No.12 (2010), art. no. 125008 .

Leahy-Hoppa, M.R.; Fitch, M.J.; Zheng, X.; Hayden, L.M. & Osiander, R. (2007). Wideband terahertz spectroscopy of explosives, *Chem. Phys. Lett.,* Vol.434, No.4 (February 2007), pp. 227-230, 0009-2614

Lewis, R.A. (2007) Physical phenomena in electronic materials in the terahertz region *Proceedings of the IEEE* Vol. 95 No. 8, art. no. 4337844 (2007), pp. 1641-1645.

Li C., Zhou M., Ding W., Du F., Li Y., Wang W .M., Sheng Z.M., Ma J.L., Chen L M, Dong Q, & Zhang J, Effects of laser plasma interactions on THz radiation from solid targets irradiated by ultra short intense laser pulses, *Physical Review E,* Vol. 84 (2011) pp. 036405-036410

Li, D. & Ma, G. (2008). Pump-wavelength dependence of terahertz radiation via optical rectification in (110)-oriented ZnTe crystal, *J. Appl. Phys.* Vol.103, No. 12, (June 2008), pp. 123101 1-4, ISSN 1089-7550

Liu, H.C., Song, C.Y., Wasilewski, Z.R., Buchanan, M. (2011) Phonon and polaron enhanced IR-THz photodetectors *Proceedings of SPIE - The International Society for Optical Engineering* 7945, (2011) art. no. 79450X

Liu, L. Xu, H.; Percy, R. R.; Herald, D. L.; Lichtenberger, A. W.; Hesler, J. L. & Weikle, R. M. (2009). Development of Integrated Terahertz Broadband Detectors Utilizing Superconducting Hot-Electron Bolometers, *IEEE trans. Appl.Supercon.* Vol.19, No. 3, (July 2009), pp. 282-286, ISSN 1051-8223

Liu, T.; Lin, G.; Lee, Y.; Wang, S.; Tani, M.; Wu, H. & Pan, C. (2005). Dark current and trailing-edge suppression in ultrafast photoconductive switches and terahertz spiral antennas fabricated on multienergy arsenic-ion-implanted GaAs, *J. Appl. Phys.* Vol.98, No.1, (July 2005), pp. 013711 1-4, ISSN 1089-7550

Lloyd-Hughes, J.; Castro-Camus, E.; Johnston, M.B. (2005). Simulation and optimisation of terahertz emission from InGaAs and InP photoconductive switches, *Solid State Commun.,* Vol.136, No.11, (October 2005) pp. 595-600, ISSN 0038-1098

Lu, X.; Karpowicz, N.; Chen, Y. & Zhang, X.-C., (2008). Systematic study of broadband terahertz gas sensor, *Appl. Phys. Lett.* Vol.93, (December 2008), pp.261106, ISSN 1077-3118

Łusakowski J., (2007), Nanometer transistors for emission and detection of THz radiation, Thin Solid Films, Vol. 515 (2007) pp. 4327–4332.

Lusakowski, J.; Knap, W.; Dyakonova, N.; Varani, L.; Mateos, J.; Gonzalez, T.; Roelens, Y.; Bollaert, S.; Cappy, A. & Karpierz, K. (2005). Terahertz emission by plasma waves in 60 nm gate high electron mobility transistors, *J. Appl. Phys.* Vol.97, No. 13 (February 2005) pp. 064307 1-3, ISSN 1077-3118

Maestrini, A.; Thomas, B.; Wang, H.; Jung, C.; Treuttel, J.; Jin, Y.; Chattopadhyay, G.; Mehdi, I. & Beaudin, G. (2010). Schottky diode-based terahertz frequency multipliers and mixers , *Comptes Rendus Physique*, Vol.11, No.7, (2010). pp. 480-495, ISSN 1631-0705

Magno, R.; Champlain, J. G.; Newman, H. S.; Ancona, M. G.; Culbertson, J. C.; Bennett, B. R.; Boos, J. B. & Park D., (2008). Antimonide-based diodes for terahertz mixers, *Appl. Phys. Lett.* Vol.92, No. 24, (June, 2008) pp. 243502 1-3, ISSN 1077-3118

Marandi, A.; Darcie, T. E.; and So, P. P. M. (2008). Design of a continuous-wave tunable terahertz source using waveguide-phase-matched GaAs, *Optics Express*, Vol.16, No.14, (June 2008) pp 10427-10433, ISSN 1094-4087

Matsuo, H. (2006). Future prospects of superconducting direct detectors in terahertz frequency range, *Nuclear Instruments and Methods in Physics Research A*, Vol.559, (January 2006), pp. 748-750, ISSN 0168-9002

McLaughlin, C. V.; Hayden, L. M.; Polishak, B.; Huang, S.; Luo, J. Kim, T. & Jen, A. K. (2008). Wideband 15 THz response using organic electro-optic polymer emitter-sensor pairs at telecommunication wavelengths, *Appl. Phys. Lett.* Vol.92, No.15, (April 2008), pp. 151107 1-3, ISSN 1077-3118

Meledin, D. V.; Marrone, D. P.; Tong, C. Y.; Gibson, .H.; Blundell, R.; Paine, S. N.; Papa, D. C.; Smith, M.; Hunter, T. R.; Battat, J.; Voronov, B. & Gol'Tsman, G. A 1-THz superconducting hot-electron-bolometer receiver for astronomical observations, *IEEE Trans. on Microwave Theory and Techniques*, Vol.52, No. 10, (October 2004), pp. 2338-2343, ISSN 0018-9480

Mendis, R.; Astley, V.; Liu, J. & Mittleman, D. M. (2009). Terahertz microfluidic sensor based on a parallel-plate waveguide resonant cavity, *Appl. Phys. Lett.* Vol.95, No.17, (October 2009), pp.171113, ISSN 1077-3118

Mittleman, D. (2003). *in Sensing with Terahertz Radiation* Springer, ISBN 3-540-43110-1, New York

Miyadera, T.; Kiwa, T.; Kawayama, I.; Murakami, Tonouchi, H. M. (2004). Ultrafast optical study of amorphous Ge thin films for superconductor/semiconductor hybrid devices, *Physica C: Superconductivity*, Vol.412, No.2 (July 2004) pp.1602-1606, ISSN 0921-4534

Molis G., Adomavicus R., Krotkus A., (2008) Temperature-dependent terahertz radiation from the surfaces of narrow-gap semiconductors illuminated by femtosecond laser pulses, *Physica B*, Vol.403 (2008) pp. 3786–3788.

Mueller E.R., (2003) Terahertz Radiation: Applications and Sources, *The Industrial Physicist*, (Aug-Sept. 2003), pp 27-29.

Nagel, M.; Först, M & Kurz, H (2006). THz biosensing devices: fundamentals and technology, *J. Phys.: Condens. Matter*, Vol.18, No.18, (April 2006), pp. S601, ISSN 1361-648X

Nagel M, Richter F., Bolivar PH, & Kurz H (2003), A functionalized THz sensor for marker free DNA analysis *Phys. Med. Biol.* Vol.48 (2003) pp.3625-36. ISSN: 0031-9155

Nikoghosyan, A.S. (2010). Laser Driven Terahertz Dielectric Wedge Antenna laced in Free Space or in Hollow Metallic Waveguide, *35th International Conference on Infrared*

Millimeter and Terahertz Waves (IRMMW-THz), (September 2010) pp. 1-2, ISBN 978-1-4244-6655-9, Rome,

Otsuji, T., Tsuda, Y., Komori, T., El Fatimy, A., Suemitsu, T. (2009) Terahertz plasmon-resonant microship emitters and their possible sensing and spectroscopic applications *Proceedings of IEEE Sensors* , art. no. 5398309, (2009) pp. 1991-1996.

Otsuji, T.; Karasawa, H.; Watanabe, T.; Suemitsu, T.; Suemitsu, M.; Sano, E. Knap, W.; Ryzhii, V. (2010). Emission of terahertz radiation from two-dimensional electron systems in semiconductor nano-heterostructures, *Comptes Rendus Physique*, Vol.11, No.7, (July 2010), pp. 421-432, ISSN 1631-0705

Peroz, C.; Degardin, A. F.; Villegier, J. C. & Kreisler, A. (2007). Fabrication and Characterization of Ultrathin PBCO/YBCO/PBCO Constrictions for Hot Electron Bolometer THz Mixing Application, *J. Appl. Supercond., IEEE Trans.*, Vol.17 , No. (July 2007), pp. 637-640, ISSN 1051-8223

Pilla, S (2007). Enhancing the photomixing efficiency of optoelectronic devices in the terahertz regime, *Appl. Phys. Lett.*, Vol.90, No.16, (April 2007), pp. 161119-22, ISSN 1077-3118

Preu, S., Dhler, G.H., Malzer, S., Wang, L.J., Gossard, A.C. (2011) Tunable, continuous-wave Terahertz photomixer sources and applications *Journal of Applied Physics* Vol.109 No.6, (2011) art. no. 061301

Putz, P.; Jacobs, K.; Justen, M.; Schomaker, F.; Schultz, M.; Wulff, S. & Honingh, C. E (2011). NbTiN Hot Electron Bolometer Waveguide Mixers on ${\rm Si}_{3}{\rm N}_{4}$ Membranes at THz Frequencies, *Appl. Superconductivity, IEEE Trans.* Vol.21, (May 2011), pp. 636-639, ISSN 1051-8223

Radhanpura, K.; Hargreaves, S.; Lewis, R. A. & Henini, M. (2009). The role of optical rectification in the generation of terahertz radiation from GaBiAs, *Appl. Phys. Lett.* Vol.94, No. 25, (June 2009), pp. 251115-1-3, ISSN 1077-3118

Radhanpura, K.; Hargreaves, S.; Lewis, R. A.; Sirbu, L. & Tiginyanu, I. M. (2010), Heavy noble gas (Kr, Xe) irradiated (111) InP nanoporous honeycomb membranes with enhanced ultrafast all-optical terahertz emission, *Appl. Phys. Lett.* Vol.97, No. 18, (November 2010), pp. 181921-1-3, ISSN 1077-3118

Rangan C. & Bucksbaum P.H., Optimally shaped THz pulses for quantum algorithm on a Rydberg atom register, *Physical Review A*, Vol. 64 (2001) pp 37402- 37410.

Reklaitis A. , (2011) Crossover between surface field and photo-Dember effect induced terahertz emission, *Journal of Applied Physics*, Vol. 109, (2011) pp. 083108(5 pages)

Reimann K., (2007), Table-top sources of ultrashort THz pulses, Reports on Progress in Physics, Vol. 70 (2007) pp.1597.

Rau, C.; Torosyan, G.; Beigang, R. & Nerkararyan, Kh., (2005). Prism coupled terahertz waveguide sensor, *Appl. Phys. Lett.* Vol.86, No.21, (2005), pp. 211119 1-3, ISSN 1077-3118

Reid, I.; Cravetchi, R.; Fedosejevs, I. M.; Tiginyanu, L.; Sirbu, & Robert W. Boyd, (2008). Enhanced nonlinear optical response of InP(100) membranes, *Phys. Rev. B* Vol.71, No. 8, (February 2005), pp. 081306, ISSN 1550-235x

Rinzan, M. B.; M. Perera, A. G. U.; Matsik, S. G.; Liu, H. C.; Wasilewski, Z. R. & Buchanan, M. (2004). AlGaAs emitter/GaAs barrier terahertz detector with a 2.3 THz threshold, *Appl. Phys. Lett.*, Vol.86, No.7 (Februvary 2005), pp. 071112 1-3, ISSN 1077-3118

Roser H.P., Hubers H.W., Crowe T.W., & Peatman W.C.B., Nanostructure GaAs Schottky diodes for far-infrared heterodyne receivers, *IR Physics* vol. 35 (1994) 451.

Roeser H.P., Haslam D.T., Hetfleisch F., Lopez J.S., vonSchoenermark M.F., Stepper M., Huber F.M., Nikoghosyan A.S., (2010) Electron transport in nanostructures: A key to high temperature superconductivity? *Acta Astronautica*, Vol. 67 (2010) pp. 546–552.

Ryzhii, V.; Satou, A.; Knap, W.; & Shur, M. S. (2006). Plasma oscillations in high-electron-mobility transistors with recessed gate, J. Appl. Phys. Vol.99, No.8, (May 2006), pp. 084507-1-5 ISSN 1089-7550

Sakai K., & Tani M., (2005) Introduction to Terahertz pulses, *Topics in Applied Physics*, Vol. 97 (2005), pp. 1-30.

Schultz, J. & Lichtenberger, A. (2007). Investigation of Novel Superconducting Hot Electron Bolometer Geometries Fabricated With Ultraviolet Lithography, *Applied Superconductivity, IEEE Trans.* Vol.17, (July 2007), No. 2, pp. 645-648, ISSN 1051-8223

Shchepetov, A.; Gardès, C.; Roelens, Y.; Cappy, A.; Bollaert, S.; Boubanga-Tombet, S.; Teppe, F.; Coquillat, D.; Nadar, S.; Dyakonova, N.; Videlier, H.; Knap, W.; Seliuta, D.; Vadoklis, R. and Valušis G. (2008). Oblique modes effect on terahertz plasma wave resonant detection in InGaAs/InAlAs multichannel transistors, *Appl. Phys. Lett.* Vol.92, No.24, (June 2008), pp. 242105-1-3, ISSN 1077-3118

Shibuya, T.; Suizu, K.; and Kawase, K.; (2010). Widely Tunable Monochromatic Cherenkov Phase-Matched Terahertz Wave Generation from Bulk Lithium Niobate, *Appl. Phys. Express*, Vol.3, (August 2010), pp. 082201-082204 ISSN 1882-0786

Sizov, F. & Rogalski A. (2010). THz detectors, *Progress in Quantum Electronics*, Vol.34, No.5, (September 2010), pp. 278-347, ISSN 0079-6727

Stern, J. A.; Bumble, B.; Kawamura, J. & Skalare, A. (2005). Fabrication of terahertz frequency phonon cooled HEB mixers, *IEEE Trans. Appl. Superconductivity*, Vol.15, No. (June 2005), pp. 499-502, ISSN 1051-8223

Suen, J. Y.; Li, W.; Taylor, Z. D. and Brown, E. R. (2010). Characterization and modeling of a terahertz photoconductive switch, *Appl. Phys., Lett.* Vol.96, No.14 (April 2010) pp. 141103-141106 ISSN 1077-3118

Suzuki, T.; Yasui, T.; Fujishima, H.; Nozokido, T.; Araki, M.; Boric-Lubecke, O.; Lubecke, V.M.; Warashina, H. & Mizuno, K. (1999). *Microwave Theory and Techniques, IEEE Trans.*, Vol.47 No.9, (September 1999), pp. 1649 – 1655, ISSN 0018-9480

Tochitsky S. Ya., Ralph J. E., Sung C., & Joshi C.(2005), Generation of megawatt-power terahertz pulses by noncollinear difference-frequency mixing in GaAs, *Journal of Applied Physics*, Vol. 98, No.2 (2005) 026101-1-3.

Tonouchi, M. (2007). Cutting edge terahertz technology, *Nature Photoic.*, Vol.1, (February 2007), pp. 97-105, ISSN 1749-4885

Tredicucci, A.; Kohler, R.; Mahler, L.; Beere, H. E.; Linfield, E. H. & Ritchie, D.A. (2005). Terahertz quantum cascade lasers—first demonstration and novel concepts, *Semicond. Sci. Technol.* Vol.20, No.7, (June 2005) pp. S222–S227, ISSN 1361-6641

Tsaur G. & Wang, J. (2009). Relativistic optical rectification driven by a high-intensity pulsed Gaussian beam, *Phys. Rev. A*, Vol.80, No.2 (August 2009) pp. 023802-1-11 ISSN 1094-1622

Turchinovich D & Dijkhuis J.I. (2007) Performance of combined <100>-<110> ZnTe crystals in an amplified THz time-domain spectrometer, *Optics Communications* Vol. 270 (2007) pp. 96–99

Ueda T & Komiyama S (2010) Novel Ultra-Sensitive Detectors in the 10-50 mm wave length range , *Sensors*, Vol. 10, (2010) 8411-8423 ISSN 1424-8220

Urbanowicz, A.; Krotkus, A.; Adomavičius, R. & Malevich, V.L. (May 2007). Terahertz emission from femtosecond laser excited Ge surfaces due to the electrical field-induced optical rectification, *Physica B: Condensed Matter*, Vol.398, No. 98 (May 2007), pp. 98-101, ISSN 0921-4526

Valavanis, A.; Dinh, T. V.; Lever, L. J. M.; Ikonić, Z. and Kelsall, R. W. (2011). Material configurations for n-type silicon-based terahertz quantum cascade lasers *Phys. Rev. B*, Vol.83, No.19, (May 2011) pp. 195321-195329 ISSN 1550-235x

Vidal, S.; Degert, J.; Tondusson, M.; Oberlé, J. & Freysz, E. (2011). Impact of dispersion, free carriers, and two-photon absorption on the generation of intense terahertz pulses in ZnTe crystals, *Appl. Phys. Lett.* Vol.98, No. 19, (May 2011), pp. 191103-1-3, ISSN 1077-3118

Videlier, H., Nadar, S., Dyakonova, N., Sakowicz, M., Trinh Van Dam, T., Teppe, F., Coquillat, D.,&Lyonnet, J., (2009), Silicon MOSFETs as room temperature terahertz detectors, *Journal of Physics: Conference Series* Vol. 193, (2009) art. no. 012095 .

Williams G.P., High-power terahertz synchrotron sources, *Philosophical Transactions of the Royal Society London* A vol. 362 (2004), pp. 403–414. ISSN: 1471-2962.

Xiao-li, Z. & Jiu-sheng, L. (2011). Diagnostic techniques of talc powder in flour based on the THz spectroscopy, *J. Phys.: Conf. Ser.* Vol.276, No.1, (March 2011) pp.012234 ISSN 1742-6596

Yang, X.; Qi, S.; Zhang, C.; Chen, K.; Liang, X.; Yang, G.; Xu, T.; Han, Y. and Tian, J. (2011), The study of self-diffraction of mercury dithizonate in polymer film, *Optics Commun.* Vol.256, No. 4, (July 2011) pp. 414-421, ISSN 0030-4018

Yasuda, H.; Kubis, T.; Vogl, P.; Sekine, N.; Hosako, I. & K. Hirakawa, (2009). Nonequilibrium Green's function calculation for four-level scheme terahertz quantum cascade lasers, *Appl. Phys. Lett.* Vol.94, No.15, (April 2009), pp. 151109 1-3, ISSN 1077-3118

Yasui, T.; Nishimura, A.; Suzuki, T.; Nakayama, K. & Okajima S. (2006). Detection system operating at up to 7 THz using quasioptics and Schottky barrier diodes, *Rev. Sci. Instrum.* Vol.77, No.6, (2006) pp-066102 1-3, ISSN 1089-7623

Yi, M., Lee, K., Lim, J., Hong, Y., Jho, Y.-D., & Ahn J. (2010) Terahertz Waves Emitted from an Optical Fiber *Optics Express* vol. 18 No. 13, (2010) pp. 13693-13699.

Yu, N.E.; Lee, K.S.; Ko, D.-K. C. l.; Kang.; Takekawa, S. l. & Kitamura, K. (2011). Temperature dependent narrow-band terahertz pulse generation in periodically poled crystals via difference frequency generation, *Optics Commun.*, Vol.284 No.5, (November 2011), pp. 1395-1400, ISSN 0030-4018

Zhang, L.; Zhong, H.; Deng, C.; Zhang, C. & Zhao, Y. (2008). Terahertz wave reference-free phase imaging for identification of explosives, *Appl. Phys. Lett.* Vol.92, No.9, (March 2008), pp. 091117, ISSN 1077-3118

Zheng, X., McLaughlin, C.V., Cunningham, P., Hayden, L.M. (2007), Organic broadband terahertz sources and sensors, *Journal of Nanoelectronics and Optoelectronics* Vol. 2 No.1 (2007), pp. 58-76.

Zheng, X., McLaughlin, C.V., Cunningham, P., Hayden, L.M. (2007), Organic broadband terahertz sources and sensors, *Journal of Nanoelectronics and Optoelectronics* Vol. 2 No.1 (2007), pp. 58-76.

Zmuidzinas J., Richards P.L., (2004), Superconducting detectors and mixers for Millimeter and sub-millimeter Astrophysics, *Proceedings of IEEE*, vol.192, No.10, (October 2004) pp 1597-1616.

6

Silicon Photomultipliers: Characterization and Applications

Marco Ramilli[1], Alessia Allevi[1], Luca Nardo[1],
Maria Bondani[2] and Massimo Caccia[1]

[1]*Dipartimento di Fisica e Matematica - Università degli Studi dell'Insubria*
[2]*Istituto di Fotonica e Nanotecnologie - Consiglio Nazionale delle Ricerche*
Italy

1. Introduction

Silicon Photo-Multipliers (SiPMs henceover) are photo-detectors based on a technology originally invented in Russia (Akindinov et al., 1997). They essentially consist of an *array of p-n junctions* operated beyond the breakdown voltage (McKay, 1954), in a Geiger-Müller (G-M) regime (Oldham et al., 1972), with typical gain of the order of 10^6 and on-cell integrated quenching mechanisms. Silicon photo-detectors with internal multiplication are in use since more than a decade (Lutz, 1995). Avalance Photo-Diodes (APDs) (Akindinov et al., 2005) are operated in a proportional regime, with typical gains of 10^4, Single-Photon Avalanche Diodes (SPADs) (Cova et al., 1996) are endowed with single-photon sensitivity and are tailored for high frequency counting with time resolutions down to 30 ps. However, being made of a single cell operated in binary mode, they do not carry any information about the intensity of the incoming light field.

SiPMs complement the family of existing sensors: with a cell density of $\sim 10^3/mm^2$, areas up to 3×3 mm^2 and *a single output node*, they offer the possibility of measuring the intensity of the light field simply by counting the number of fired cells. The main features of SiPMs, due to their structural and operational characteristics are:

- high gain, granted by G-M operating mode, comparable to the values achieved by standard photomultipliers (PMTs);
- enhanced linearity, owing to cell structure, with deviations that become relevant when the average number of detected photons approaches the same order of magnitude of the number of cells of the device (Tarolli et al., 2010);
- large dynamic range, provided by the pixelated structure with a common output, spanning from the single photon regime up to high intensities;
- operability in magnetic fields, compactness and relatively low cost, granted by silicon-based technology.

On the other hand, since G-M avalanches that are triggered by electron-hole pairs extracted by the impinging photons are obviously indistinguishable from the ones originated from other processes, dealing with all possible sources of noise is far from being trivial:

- thermally extracted electron-hole pairs cause high Dark Count Rates (DCR), with values ranging from several hundreds of kHz up to the MHz level, depending on the total number of cells, the operating temperature and the overbias;
- spurious signals are also due to optical cross-talk (Sciacca et al., 2008): photons emitted by an avalanche can travel through silicon and reach the depleted region in a neighboring cell, thus triggering another avalanche;
- carriers extracted during an avalanche process may be trapped in a false potential minimum in the depleted region: escaping from that trap they can originate an *afterpulse* avalanche, so called because it typically happens shortly after (or even during) the recovery of the previous avalanche (Du et al., 2008; Eckert et al., 2010).

These apparently huge disadvantages can be overridden by means of an exhaustive characterization of the SiPM performances. An example of this procedure will be given in the following Section, where we present the characterization of the devices as a function of temperature.

Another aspect of the characterization of SiPMs will be addressed by providing a description of the G-M avalanches probability distribution functions, in order to reconstruct the statistics of the impinging light (Ramilli et al., 2010).

Modeling the SiPM response is also preliminary to application of these sensors as detectors in Fluorescence Fluctuation Spectroscopy biophysical experiments (Chen et al., 1999; Schwille, 2001), where the parameters describing the system under investigation are inferred by the deviations of the fluorescence intensity of suitable probes around its mean value. A feasibility study on this topic will be presented in the final part of the Chapter.

2. Characterization

Tests have been performed on existing devices with the main goal to define an exhaustive characterization protocol and to produce a comparative study. Three kinds of detectors from different manufacturers have been studied: SensL[1], Hamamatsu Photonics[2] and STMicroelectronics[3]. Detector characterization is a major task for all the applications of SiPMs and in particular for the identification of critical parameters: for example, DCR is an important parameter for low-rate-event applications, while thermal stability is essential for portable devices. For this reason, the following characterization protocol has been developed assessing:

- Geometrical parameters (number of cells, size of detectors and occupancy factor);
- I–V measurements;
- Noise measurements: DCR, optical cross-talk, dependence on the environmental parameters.
- Analysis of photon spectra: resolution power, gain, working point optimization (at low and large flux), electronic noise measurement taking into account cell-to-cell variations, dependence on the environmental parameters (temperature).
- Linearity and dynamic range.
- Spectral response measurement: photon detection efficiency (PDE).

---

[1] see http://sensl.com/.
[2] see http://www.hamamatsu.com/.
[3] see http://www.st.com/stonline/.

In particular, studies of the main SiPM parameters (such as gain, PDE, DCR and optical cross-talk) as functions of temperature will be described in details.

2.1 Experimental setup

In order to implement the characterization protocol, we have developed an experimental setup in which a green-emitting LED (λ = 510 nm), coupled to a fast pulse generator (PDL800-B PicoQuant), has been used as the light source. The SiPM output signal has been elaborated in the following way:

- SiPM was directly connected to a first stage amplification board:
    - in experimental situations where the SiPM was operated under a continuous light flux, a transimpedance pre-amplifier provided by SensL has been used: this device converts the raw current from the SiPM into a voltage, with an amplification of 470 V/A;
    - when the SiPM was operated in the pulsed regime, a different pre-amplifier, called Pulse Amplifier and also provided by SensL, has been used. It allowed the fast rise of the detector to be exploited, providing an amplification factor of 20;
- a leading edge discriminator (Lecroy 821) with a user-defined voltage threshold that the SiPM output has to exceed in order to provide a triggering signal has been then exploited at the output of the amplifier;
- in case of frequency measurements (e.g. DCR measurements), the discriminator output has been directly sent to a scaler;
- the output of the SiPM has been integrated by a CAEN QDC V792N board:
    - the board provides a charge measurement performing the integration of the input voltage signal, with a conversion time of 2.8 μs and a reset time of 4 μs;
    - the integration gate is generated by a NIM timing unit and must precede the analog input signal of at least 15 ns;
- data have been stored in a PC via a USB-VME Bridge (CAEN).

The system composed by the investigated SiPM and the first-stage amplification board has been located in a metal box sealed with grease, in which air has been replaced with helium. The cooling fluid has been pumped into the box through a copper pipe, allowing controlled temperature variations. The temperature of the system has been measured by a thermistor placed in contact with the external packaging of the sensor.

2.2 Studying temperature behavior

In order to quantitatively understand the effects of temperature changes on the SiPM main figures of merit, a suitable set of measurements must be accomplished. The procedure has been tested with SensL CSI 0747 015 A20 HD and the results regarding this sensor will be presented in more detail in the following. Once the procedure was tested, it has been repeated with different SiPM models: a SensL CSI 0740 001 A20 HD, a Hamamatsu S10362-11-100C and a ST Microelectronics TO-8 prototype.

Sensors have been placed in the cooling box described in Section 2.1 and several spectra have been acquired at different bias voltages and different temperatures in order to obtain the values of gain and PDE. Moreover, DCR measurements at different thresholds were used to evaluate both DCR and cross-talk contributions. To analyze the procedure in detail, a typical

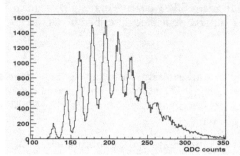

Fig. 1. Example of a low flux spectrum of a SiPM; each histogram bin represents a single QDC channel corresponding to 0.11 pC.

low-flux spectrum is presented in Fig. 1 that shows the charge measurements performed by the CAEN QDC: a waveform generator triggered both the light source and the *gate* needed to perform the charge measurement. The structure of the histogram reflects the characteristics of the SiPM, since the n-th peak position represents the most probable output value (in released charge) of n cells firing simultaneously and its Gaussian broadening is due to stochastic noise sources. In this experimental situation, the very first peak represents the *zero-photon* peak, i.e. the output of the QDC board with no SiPM signal. The broadening of the n-th peak σ_n can be described with good approximation as:

$$\sigma_n^2 = \sigma_0^2 + n\sigma_1^2, \tag{1}$$

where σ_0 is the variance of the zero-photon peak (giving the noise contribution due to the electronic chain), and σ_1 is the variance of the first photon peak, providing an estimation of the SiPM noise.

2.2.1 Gain

The gain has been evaluated by illuminating the SiPM with a low photon flux, thus obtaining a spectrum whose peaks are clearly recognizable (see Fig. 1). As the n-th peak corresponds to the mean charge released by n G-M avalanches, the gain G can be computed in the following way:

$$G = \frac{QDC_{cal}}{e^- K_{amp}} \Delta_{PP}, \tag{2}$$

where $QDC_{cal} = 0.11$ pC/channel is the charge corresponding to one QDC unit, e^- is the elementary charge, K_{amp} is the global amplification factor of the electronic setup, and Δ_{PP} is the distance in QDC units between two adjacent peaks of the collected spectra.

The gain behavior as a function of the bias voltage, at fixed temperatures, is shown in Fig. 2: a linear dependence in the range of interest is clearly observable with a slope independent of the temperature within the experimental errors. In Fig. 3 the same gain values are presented as functions of temperature, for fixed bias voltages: a linear behavior is still clearly recognizable; in this case as well the slope is not affected by the change of the applied bias.

This analysis suggests that the gain can be expressed as a linear function of a variable which can be re-scaled with temperature. Since it is well known that the breakdown voltage has, in

SiPM Model	m_{BD} (mV/degree)
SensL CSI 0747 015 A20 HD	23.2 ± 1.4
SensL CSI 0740 001 A20 HD	23.6 ± 0.9
Hamamatsu S10362-11-100C	61.9 ± 0.7
STM TO-8 prototype	31.5 ± 0.1

Table 1. Rate of change of the breakdown voltage with temperature.

our range of interest, a linear dependence on temperature (Goetzberger et al., 1963), the over voltage, defined as the difference between the applied bias voltage and the breakdown one, is thus a suitable candidate. The conditions can be summarized as follows

$$G(V, T) = m_V(V - V_{BD}(T)), \tag{3}$$

$$G(V, T) = m_T T + G(T_0, V), \tag{4}$$

where m_V and m_T are the slopes, $V_{BD}(T)$ is the breakdown voltage and T_0 is a reference temperature. Solving these equations for $V_{BD}(T)$, the rate of change of the breakdown voltage with temperature can be expressed as:

$$m_{BD} = -\frac{m_T}{m_V}. \tag{5}$$

The procedure has been applied to the different SiPMs. From the results in Table 1, we can see that the values of m_{BD} are technology dependent.

By using the breakdown voltage value measured at room temperature (RT) as the reference

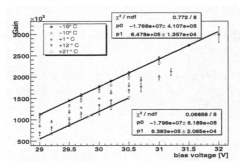

Fig. 2. Gain as a function of bias voltage, for different temperatures; a linear fit has been performed for each temperature set: the obtained fit slopes are in agreement.

value, the breakdown values for each temperature can be written as:

$$V_{BD}(T) = m_{BD}(T - T_{RT}) + V_{BD}(T_{RT}). \tag{6}$$

From Equation 6 it is possible to express the gain and all the other measured parameters as a function of the over voltage: in Fig. 4 the gain is plotted as a function of the over voltage, showing a global linear behavior independent of the temperature. Equation 6 can be used to fix the operational parameters of a SiPM regardless of the varying environmental conditions: if m_{BD} is known with enough precision and frequent measurements of the SiPM temperature are performed, the over voltage across each cell can be maintained fixed by simply continuously adjusting the applied bias according to the temperature variations. This

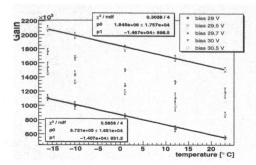

Fig. 3. Gain as a function of temperature (expressed in Celsius degrees), for fixed bias voltages; each set has been fitted using a linear law: the obtained slopes are in agreement.

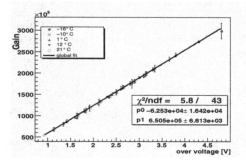

Fig. 4. Gain as a function of the over voltage: values are obtained by calculating the breakdown voltage for each temperature; all the data acquired at different temperatures have been fitted with the same linear law.

idea has led to a collaboration between Università dell'Insubria and CAEN for the realization of CAEN SP5600 General Purpose Power Supply and Amplification Unit module with an integrated threshold discriminator.

2.2.2 Dark count rate

As explained in Section 1, DCR is the frequency of the G-M avalanches triggered by thermally extracted carriers. A scan of the DCR at different thresholds can be done, and the resulting plot (an example is shown in Fig. 5) is usually referred to as *staircase function*.
DCR has been measured at different voltages and different temperatures, also setting different discrimination thresholds; in Fig. 6, for example, the results for a *"half photon threshold"* are shown. A clear and expected dependence on temperature is recognizable.
An exhaustive knowledge of DCR behavior is of utmost importance for a complete characterization of all the noise sources of the detector. It can be a fundamental figure of merit for low flux applications.

2.2.3 Optical cross-talk

An electron avalanche is modeled as a microplasma (Oldham et al., 1972): photons emitted by the accelerated carriers during this event have a certain probability to reach the neighboring

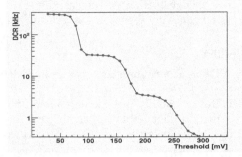

Fig. 5. Example of *staircase curve*, obtained after a three-stage amplification.

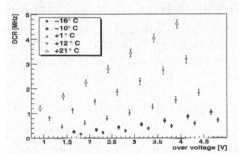

Fig. 6. DCR at different temperatures as a function of the over voltage, for a SensL SiPM.

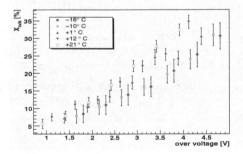

Fig. 7. Cross-talk as a function of the over voltage, at different temperatures, for a SensL SiPM.

cell diodes thus triggering a second avalanche. The quantity named optical cross-talk (in short cross-talk, X_T) is simply the percentage of avalanches triggered by such a mechanism.

Cross-talk has been calculated starting from DCR measurements by taking the ratio between the DCR frequencies with the discriminator threshold respectively at *"one-and-half photon"* and at *"half photon"*. This method is based on the assumption that the probability that two uncorrelated thermally-triggered avalanches are generated within the same rise time is negligible, so that all the second photon events are due to cross-talk.

In Fig. 7 the results of the cross-talk evaluation are presented: it is worth noting that, within the experimental errors, cross-talk does not seem to suffer a strong temperature dependence.

2.2.4 Photon detection efficiency

Photon detection efficiency (PDE) is a key parameter for every light detector and is defined as the product of three terms:
- quantum efficiency, that is the probability that the impinging photon produces a charge carrier, which is a function of the incoming light wavelength;
- the G-M probability that the extracted carrier generates an avalanche that depends on the over voltage;
- the so-called Fill Factor (FF), a geometrical factor that expresses the portion of the active area of the sensor with respect to the total area.
The PDE value thus represents the fraction of impinging photons that are actually detected.

To measure this quantity, SiPMs have been illuminated with a light intensity Φ which has been previously measured with a calibrated PMT HAMAMATSU H5783 and the resulting spectra have been acquired. PDE has been evaluated by estimating the mean number of detected photons $< n >_{meas}$ with respect to the incoming light intensity:

$$PDE = \frac{< n >_{meas}}{\Phi} \tag{7}$$

In the case of spectra in which each photon peak could be resolved, $< n >_{meas}$ has been obtained by fitting the peak positions with a Poissonian curve and by evaluating the mean value of the fit distribution. In the case in which the peaks could not be resolved, $< n >_{meas}$ has been estimated as

$$< n >_{meas} = \frac{QDC_{cal}}{e^- \, G \, K_{amp}} \Delta QDC \tag{8}$$

where G is the proper gain value, e^- is the elementary charge, $QDC_{cal} = 0.11$ pC is the charge per QDC unit, K_{amp} is the amplification factor of the electronics chain, and ΔQDC is the difference between the mean value of the obtained spectrum and the pedestal position.

However, this method provides a zero-order approximation of the PDE because it does not take into account cross-talk effects that can be as large as 40% (see Fig. 7 as a reference). To properly estimate the "true" number $< n >_0$ of impinging photons triggering an avalanche, the following relation has been used:

$$< n >_{meas} = \frac{< n >_o}{1 - X_T} \simeq < n >_o (1 + X_T). \tag{9}$$

The approximation is valid for small values of the cross-talk; the previously obtained PDE values have thus been corrected using $< n >_0$ as the mean number of detected photons as shown in Fig. 8. Again, PDE does not seem to have any remarkable temperature dependance, thus confirming the agreement of our results with the definition of PDE.

3. Photon-number statistics

Due to their good linear response, SiPMs can be considered as the ideal candidates for the reconstruction of the statistics of any light state (Afek et al., 2009). Obviously, the presence of DCR and cross-talk effects must be taken into account. A proper model of the detector

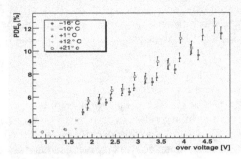

Fig. 8. PDE values corrected taking into account cross-talk effects, for a SensL SiPM. The wavelength of the impinging light is 510 nm.

	Hamamatsu MPPC S10362-11-100C
Number of Diodes:	100
Area:	1 mm × 1 mm
Diode dimension:	100 μm × 100 μm
Breakdown Voltage:	69.23 V
Dark Count Rate:	540 kHz at 70 V
Optical Cross-Talk:	25 % at 70 V
Gain:	3.3 · 10^6 at 70 V
PDE (green):	15 % at 70 V

Table 2. Main features of the SiPM MPPC S10362-11-100C (Hamamatsu). The data refer to room temperature.

response is thus required in order to properly assess the statistical properties of the light field under investigation (Ramilli et al., 2010). In the following Section we present two different analysis methods, both validated by suitable measurements performed by using a SiPM produced by Hamamatsu Photonics. In particular, to keep DCR and cross-talk effects at reasonable levels and to optimize the PDE, a detector endowed with 100 cells was chosen (MPPC S10362-11-100C, see Table 2). To test the model, the sensor was illuminated by the pseudo-thermal light obtained by passing the second-harmonics (@ 523 nm, 5.4-ps pulse duration) of a mode-locked Nd:YLF laser amplified at 500 Hz (High Q Laser Production) through a rotating ground-glass diffuser (D in Fig.8). The light to be measured was delivered to the sensor by a multimode optical fiber (1 mm core diameter). The detector response was integrated by the charge digitizer V792 by Caen. The signal was typically integrated over a 200-ns-long time window synchronized with the laser pulse.

3.1 Detector response modeling

The response of an ideal detector to a light field can be described as a Bernoullian process (Mandel & Wolf, 1995):

$$B_{m,n}(\eta) = \binom{n}{m} \eta^m (1 - \eta)^{n-m} \tag{10}$$

being n the number of impinging photons over the integration time, m the number of detected photons and $\eta < 1$ the PDE. Actually, η is a single parameter quantifying detector effects and losses (intentional or accidental) due to the optical system. As a consequence, the distribution $P_{m,\text{det}}$ of the number of detected photons has to be linked to the distribution $P_{n,\text{ph}}$ of the number of photons in the light under measurement by (Agliati et al., 2005; Mandel & Wolf, 1995; Zambra et al., 2004):

$$P_{m,\text{det}} = \sum_{n=m}^{\infty} B_{m,n}(\eta) P_{n,\text{ph}} \tag{11}$$

It can be demonstrated (Bondani et al., 2009a) that for a combination of classical light states the statistics is preserved by the primary detection process. This simple description has to be further developed to link $P_{m,\text{det}}$ to the probability density distribution of the G-M avalanches of any origin. First we must take into account spurious hits and cross-talk effects, not negligible in the detectors under study. The DCR results in a Poissonian process that can be described as:

$$P_{m,\text{dc}} = \overline{m}_{\text{dc}}^m / m! \exp(-\overline{m}_{\text{dc}}) \tag{12}$$

where $\overline{m}_{\text{dc}}$ is the mean number of dark counts during the gate window and $\sigma_{m,\text{dc}}^{(2)} = \sigma_{m,\text{dc}}^{(3)} = \overline{m}_{\text{dc}}$. As a consequence, the statistics of the recorded pulses may be described as:

$$P_{m,\text{det}+\text{dc}} = \sum_{i=0}^{m} P_{i,\text{dc}} P_{m-i,\text{det}} \tag{13}$$

obviously shifting the mean value and increasing variance and third-order central moment in the following way: $\overline{m}_{\text{det}+\text{dc}} = \overline{m}_{\text{det}} + \overline{m}_{\text{dc}}$, $\sigma_{m,\text{det}+\text{dc}}^{(2)} = \sigma_{m,\text{det}}^{(2)} + \overline{m}_{\text{dc}}$ and $\sigma_{m,\text{det}+\text{dc}}^{(3)} = \sigma_{m,\text{det}}^{(3)} + \overline{m}_{\text{dc}}$. As a further step, cross-talk effects must be taken into account. Cross-talk is a genuine cascade phenomenon that can be described at first order as (Afek et al., 2009)

$$C_{k,l}(x_t) = \binom{l}{k-1} x_t^{k-l}(1 - x_t)^{2l-k}. \tag{14}$$

being x_t the (constant) probability that the G-M avalanche of a cell triggers a second cell (which becomes equivalent to the cross-talk probability X_T in the limit of $X_T \to 0$), l the number of dark counts and photo-triggered avalanches and k the actual light signal amplitude. Within this first-order approximation, the real sensor response is described by

$$P_{k,\text{cross}} = \sum_{m=0}^{k} C_{k,m}(x_t) P_{m,\text{det}+\text{dc}} \tag{15}$$

characterized by $\overline{k}_{\text{cross}} = (1 + x_t)\overline{m}_{\text{det}+\text{dc}}$, $\sigma_{k,\text{cross}}^{(2)} = (1 + x_t)^2 \sigma_{m,\text{det}+\text{dc}}^{(2)} + x_t(1 - x_t)\overline{m}_{m,\text{det}+\text{dc}}$ and $\sigma_{k,\text{cross}}^{(3)} = (1 + x_t)^3 \sigma_{m,\text{det}+\text{dc}}^{(3)} + 3x_t(1 - x_t^2)\sigma_{m,\text{det}+\text{dc}}^{(2)} + x_t(1 - 3x_t + 2x_t^2)\overline{m}_{m,\text{det}+\text{dc}}$.

In the following we refer to this analytical model as *Method I* and compare it with a better refined model (*Method II*). This second method offers, in principle, an extended range of application even if it is limited to a numerical rather than an analytical solution. Irrespective of the model, the amplification and digitization processes that produce the output x can be simply described as a multiplicative parameter G:

$$P_{x,\text{out}} = G P_{Gk,\text{cross}} \tag{16}$$

whose moments are $\overline{x}_{\text{out}} = G\overline{k}_{\text{cross}}$, $\sigma_{x,\text{out}}^{(2)} = G^2 \sigma_{k,\text{cross}}^{(2)}$ and $\sigma_{x,\text{out}}^{(3)} = G^3 \sigma_{k,\text{cross}}^{(3)}$.

3.1.1 Method I: an analytical evaluation of the second and third order momenta

In Ref. (Bondani et al., 2009b) we presented a self-consistent method aimed at reconstructing the statistics of detected photons. According to this method, the analysis of the output of the detector is based on the assumptions that the detection process is described by a Bernoullian convolution and that the overall amplification-conversion process is given by a very precise constant factor G, which allows the shot-by-shot detector output to be converted into the number of detected photons. Experimentally, the value of G can be obtained by detecting a light field at different optical losses and keeping the detector parameters fixed. Once the amplification-conversion factor has been evaluated, the detected-photon distribution can be achieved by dividing the output values by G and re-binning the data in unitary bins. Here we present the extension of the method to detectors with a significant DCR and first order cross-talk effects. The second-order momentum of the recorded pulse distribution $P_{x,\text{out}}$ can be used to evaluate the Fano factor:

$$F_{x,\text{out}} = \frac{\sigma_{x,\text{out}}^{(2)}}{\overline{x}_{\text{out}}} = \frac{Q_{\text{det}+\text{dc}}}{\overline{m}_{\text{det}+\text{dc}}} \overline{x}_{\text{out}} + G \frac{1 + 3x_t}{1 + x_t} , \qquad (17)$$

where $Q_{\text{det}+\text{dc}} = \sigma_{m,\text{det}+\text{dc}}^{(2)}/\overline{m}_{\text{det}+\text{dc}} - 1$ is the Mandel factor of the primary charges. Note that, due to dark counts, the coefficient of $\overline{x}_{\text{out}}$ in Equation 17 cannot be written as $Q_{\text{ph}}/\overline{n}$ (Andreoni & Bondani, 2009; Bondani et al., 2009b), that is, the coefficient $Q_{\text{det}+\text{dc}}/\overline{m}_{\text{det}+\text{dc}}$ does not only depend on the light field to be measured. Similarly we can calculate a sort of symmetry parameter

$$S_{x,\text{out}} = \frac{\sigma_{x,\text{out}}^{(3)}}{\overline{x}_{\text{out}}} = \frac{Q_{s,\text{det}+\text{dc}} - 3Q_{\text{det}+\text{dc}}}{\overline{m}_{\text{det}+\text{dc}}^2} \overline{x}_{\text{out}}^2 + G \frac{1 + 3x_t}{1 + x_t} \frac{Q_{\text{det}+\text{dc}}}{\overline{m}_{\text{det}+\text{dc}}} \overline{x}_{\text{out}} + G^2 \frac{1 + 7x_t}{1 + x_t} , \qquad (18)$$

where $Q_{s,\text{det}+\text{dc}} = \sigma_{m,\text{det}+\text{dc}}^{(3)}/\overline{m}_{\text{det}+\text{dc}} - 1$. In the presence of dark counts both coefficients $(Q_{s,\text{det}+\text{dc}} - 3Q_{\text{det}+\text{dc}})/\overline{m}_{\text{det}+\text{dc}}^2$ and $Q_{\text{det}+\text{dc}}/\overline{m}_{\text{det}+\text{dc}}$ depend on the light being measured (Bondani et al., 2009a).

3.1.2 Method II: a numerical evaluation based on the photon-number resolving properties of SiPMs

The above mentioned self-consistent method is very powerful, but requires the acquisition of several histograms at varying η, which might not always be easily performed in practical applications: from this point of view, the possibility to analyze each spectrum independently looks complementary to the self-consistent approach. This analysis has been performed with a two-step procedure:

- the areas of the spectrum peaks have been computed in order to obtain an estimation of the number of counts per peak;
- the obtained data points have been fitted with a theoretical function, which takes into account the statistics of the light, the detection and all the deviations of the detectors from ideality, such as DCR and cross-talk effects.

To evaluate the area of each peak, we performed a multi-peak fit of the spectrum histogram modeling each peak with a Gauss-Hermite function (Van Der Marel et al., 1993):

$$\text{GH} = Ne^{-w^2/2} \left[1 + {}^3hH_3(w) + {}^4hH_4(w) \right] \qquad (19)$$

where $w = (x - \bar{x})/\sigma$, N is a normalization factor, $\bar{x}$ the peak position and σ the variance of the Gaussian function. $H_3(w)$ and $H_4(w)$ are the third and the fourth normalized Hermite polynomials and their contribution gives the asymmetry of the peak shape, whose entity is regulated by the pre-factors ^3h and ^4h, with values in the range $[-1, 1]$. The global fit function of the spectrum is a sum of as many Gauss-Hermite functions as the number of resolved peaks.

By using Equation 19 it is possible to calculate the area A_n of the n-th peak as:

$$A_n = N_n \sigma_n \left(\sqrt{2\pi} + {}^4h_n\right). \tag{20}$$

The error σ_{A_n} has been calculated by propagating the errors on the fit parameters.
This analysis is also useful to estimate the system gain G: in fact, from the fitted values of the peak positions $\bar{x}_n$, the peak-to-peak distance Δ for all the resolved peaks is:

$$\Delta_{n,n+1} = \bar{x}_{n+1} - \bar{x}_n. \tag{21}$$

The associated error $\sigma_{\Delta_{n,n+1}}$ can be once again obtained by propagating the fit errors; furthermore, the value of G can be derived as a weighted average on all the peak-to-peak values obtained from the analysis of the histograms.

The effect of detection, DCR and amplification have been modeled as described in the previous Sections (see Equations 10-13 and Equation 16).

The effect of cross-talk has been described by using a Bernoullian process, in a way analogue to what has been done in Equation 14. However, as the cross-talk process is intrinsically a cascade phenomenon, its contribution has been calculated by adding higher order effects:

$$P_{k,cross} = \sum_{m=0}^{k} \sum_{n=0}^{m} \sum_{j=0}^{n} P_{k-m-n-j,det+dc} \, B_{m,k-m-n-j}(x_t) \, B_{n,m}(x_t) \, B_{j,n}(x_t); \tag{22}$$

where the terms like $B_{j,n}(x_t)$ stand for the Bernoullian distribution

$$B_{j,n}(x_t) = \binom{n}{j} x_t^j (1 - x_t)^{n-j}. \tag{23}$$

Such a higher order expansion is not trivial to be achieved by the self-consistent approach of *Method I*, in which an explicit analytic expression of $P_{x,out}$ is needed in order to calculate its moments. Here, as all the elements of interest ($\bar{m}_{el}$, $\bar{m}_{dc}$, x_t, the number of modes μ) will be obtained as parameters of a fit, this is not necessary and therefore $P_{x,out}$ can be just numerically computed as the fitting function.

The major limit of this approach is obvious: as all the information on the statistics of the system is obtained from the peak areas, this method can only be applied to peak-resolving histograms with a number of peaks larger than the number of free parameters of the fitting function, which, in the present analysis, can rise up to five.

3.2 Experimental results

In Ref. (Ramilli et al., 2010), we presented the experimental validation of the two complementary methods by using both coherent and pseudo-thermal light. In particular, we have demonstrated that it is not possible to use coherent light to derive the DCR contribution,

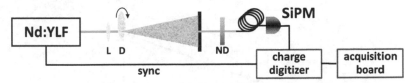

Fig. 9. Experimental setup. Nd:YLF: laser source, ND: variable neutral density filter, SiPM: detector.

which is described by a Poissonian distribution and is thus indistinguishable from the light under study. So we think that, within this context, it is more exhaustive and intriguing to show the results obtained with thermal light. The photon-number distribution of a field made of μ independent thermal modes is given by:

$$P_{n,\text{ph}} = \frac{(n+\mu-1)!}{n!\,(\mu-1)!\,(N_{th}/\mu+1)^{\mu}\,(\mu/N_{th}+1)^{n}},\tag{24}$$

with $\bar{n} = N_{th}$, $\sigma_n^{(2)} = N_{th}\,(N_{th}/\mu+1)$ and $\sigma_n^{(3)} = N_{th}\,(N_{th}/\mu+1)\,(2N_{th}/\mu+1)$.

We reproduced a single mode of this pseudo-thermal light field by passing the coherent light of our laser source through an inhomogeneous diffusing medium and then selecting a single speckle with an aperture ($\sim 150\ \mu m$ diameter), much smaller than the coherence area of the speckle pattern produced (Arecchi, 1965). By delivering this pseudo-thermal light to the SiPM sensor, we measured the values of the output, x, at 50000 subsequent laser shots and at 10 different mean values, obtained by means of a variable neutral-density filter (ND in Fig. 9). In Fig. 10 the plot of the experimental values of $F_{x,\text{out}}$ and $S_{x,\text{out}}$ is shown as a function of $\bar{x}_{\text{out}}$, along with the fitting curves evaluated from Equations 17 and 18:

$$F_{x,\text{out}} = \left(1 - \frac{\bar{x}_{\text{dc}}}{\bar{x}_{\text{out}}}\right)^2 \bar{x}_{\text{out}} + B$$

$$S_{x,\text{out}} = A\left(1 - \frac{\bar{x}_{\text{dc}}}{\bar{x}_{\text{out}}}\right)^3 \bar{x}_{\text{out}}^2 + 3B\left(1 - \frac{\bar{x}_{\text{dc}}}{\bar{x}_{\text{out}}}\right)^2 \bar{x}_{\text{out}} + C,\tag{25}$$

where μ, the number of independent thermal modes, has been set equal to one and, for simplicity of notation, the parameters $A = 2$, $B = G(1+3x_t)/(1+x_t)$ and $C = G^2(1+7x_t)/(1+x_t)$ have been introduced. We fitted the data to $F_{x,\text{out}}$ thus obtaining the values of $\bar{x}_{\text{dc}} = (5.82028 \pm 1.34015)$ and $B = (87.805 \pm 2.09009)$ ch.

Then the data for $S_{x,\text{out}}$ have been fitted substituting the above values of $\bar{x}_{\text{dc}}$ ch and B in order to obtain $A = (2.34754 \pm 0.091576)$ and $C = (8531.48 \pm 419.571)$ ch^2. These values were then used to evaluate G and x_t, obtaining $G = (74.2785 \pm 1.76982)$ ch and $x_t = (0.100174 \pm 0.050529)$. The x-values were then divided by G and re-binned in unitary bins (Bondani et al., 2009b) to obtain the $P_{k,\text{cross}}$ distribution of the actual light signal amplitude measured in the presence of dark counts and cross-talk. In Fig. 11 six different $P_{k,\text{cross}}$ distributions are plotted as bars at different mean values. Superimposed to the experimental values, two theoretical distributions are shown: the first one (white circles) is evaluated including the dark-count contribution that modifies the statistics of a single-mode thermal

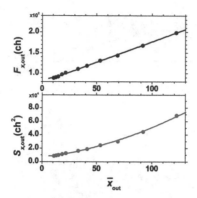

Fig. 10. Plot of $F_{x,\text{out}}$ and $S_{x,\text{out}}$ as functions of $\bar{x}_{\text{out}}$ for pseudo-thermal light.

distribution (see Equation 24) into

$$
P_{m,\text{det}+\text{dc}} = \sum_{k=0}^{m} P_{k,\text{dc}} P_{m-k,\text{det}} =
$$
$$
= \frac{e^{-\bar{m}_{\text{dc}}}}{(\mu - 1)!} \left(1 + \frac{\mu}{\bar{m}_{\text{det}}}\right)^{-m} \left(1 + \frac{\bar{m}_{\text{det}}}{\mu}\right)^{-\mu}
$$
$$
U\left[-m, 1 - m - \mu, \bar{m}_{\text{dc}}\left(1 + \frac{\mu}{\bar{m}_{\text{det}}}\right)\right], \qquad (26)
$$

where $U(a, b, z)$ is the confluent hypergeometric function. The parameters are evaluated as $\bar{m}_{\text{dc}} = \bar{x}_{\text{dc}}/(G(1 + x_t))$ and $\bar{m}_{\text{det}} = (\bar{x}_{\text{out}} - \bar{x}_{\text{dc}})/(G(1 + x_t))$. The second curve (full circles) is evaluated from Equation 15 to take into account the cross-talk. Unfortunately, the calculation does not yield an easy analytical result, and hence it has been evaluated numerically.

The comparison between the data and the theoretical functions can be estimated through the evaluation of the fidelity $f = \sum_{k=0}^{m} \sqrt{P_{k,\text{exp}} P_{k,\text{theo}}}$ (see Fig. 11).

Turning now to the other approach, it is worth noting that the number of fit parameters is large: the probability distribution is described by the expectation value $\bar{m}_{\text{det}}$ of the avalanches generated by detection, the expectation value $\bar{m}_{\text{dc}}$ of DCR contribution, the number of modes μ, the probability x_t of triggering a cross-talk event (up to three "iterations") and again a global normalization factor, for a total of 5 fit parameters: obviously, this puts a severe limit on the applicability of this method, needing at least 6 resolved peaks.

As it can be noted from the fit results in Fig. 12, the results obtained by using *Method II* are compatible within errors with what we found by applying *Method I*. However, even if the global fits present a very low χ^2 value for degree of freedom, the obtained fit parameters present large uncertainties, probably indicating the presence of very large off-diagonal elements in the minimization matrix and suggesting a strong correlation between the various parameters. This problem can be avoided by fixing some of the fit parameters (such as DCR or cross-talk), once their values have been retrieved from an accurate direct measurement (see Section 2).

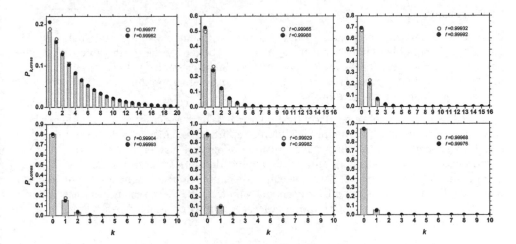

Fig. 11. Experimental $P_{k,cross}$ distributions at different mean values (bars) and theoretical distributions evaluated according to *Method I*: thermal modified by dark count distribution (white circles), thermal modified by dark counts and cross-talk effect (full circles). The corresponding fidelity values of the reconstruction are also shown.

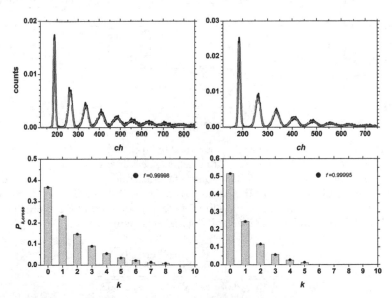

Fig. 12. Experimental results for *Method II* applied on two of the histograms acquired with thermal light. Upper panels: result of the multi-peak fit procedure; lower panels: fitted theoretical function. The corresponding fidelity values of the reconstruction are also shown.

3.3 Discussion

It is worth comparing the results of the two analysis methods on the same data sets (see Fig. 13). It is possible to notice that the agreement is better for the mean values of detected photons (panel (a)) and for the DCR (panel (b)), while the estimated values of x_t with the two *Methods* definitely disagree. This can be due to the different approximations adopted by the two *Methods* (first order *vs* third order).

In Table 3 we summarize the results of the two *Methods*. Both *Methods* work in a self-consistent way, even if they have two definitely different approaches. *Method I* does not need peak resolving capability, but requires the acquisition of several histograms at varying η. Once determined the parameters x_t and DCR, all the datasets in a series can be analyzed, independent of the number of distinguishable peaks in the pulse-height spectrum. *Method II* works analyzing each histogram independently, but, as the G-M-avalanches distribution is obtained with a fit of the data, it requires at least a number of resolved peaks larger than the number of free parameters.

The fact that the two *Methods* give very similar results for mean photon numbers is particularly important as in most applications this is the only important parameter. Merging the two *Methods* we can develop an optimal strategy based on a self-consistent calibration performed by measuring a known light and analyzing the data with *Method I*: once x_t and DCR are known, the determination of the mean photon number is independent of the specific light statistics. Hence the information on the mean photon number can be obtained from each single measurement, even when the fitting procedure of *Method II* cannot be applied.

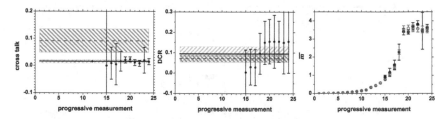

Fig. 13. Left panel: Cross- talk values of x_t obtained by applying *Method II* to thermal light (full circles) and their weighted average (full line). Dashed line: value of x_t obtained by *Method I*. Central panel: Values of DCR evaluated for the same data as in the left panel by applying *Method II* (full circles) and their weighted average (full line). Dashed line: value of DCR obtained by *Method I*. Right panel: Values of mean photon numbers evaluated by applying *Method II* (full circles) and by applying *Method I* (white circles).

	Pseudo-thermal	
	Method I	*Method II*
DCR	0.071 ± 0.017	0.094 ± 0.035
Cross-talk	0.091 ± 0.042	0.035 ± 0.004

Table 3. Comparison between the global DCR and cross-talk values obtained with *Method I* and weighted average of the values obtained with *Method II*.

4. Photon counting histogram with SiPM

Another instance in which the capability of resolving pulse-by-pulse the number of photons in high-repetition-rate periodic light signals is the analysis of fluorescence fluctuations. Traditionally, fluorescence techniques are applied to the study of systems constituted by huge ensembles of particles. For this reason, the probability distribution of the photon emission rate is peaked around its average value and fluctuations in the fluorescence signal are negligibly small. Nevertheless, if a microscopic system in which only few fluorophores are excited at a time is considered, the fluctuations in the fluorescence intensity become comparable to the average fluorescence intensity value, and can thus be measured. As these fluctuations are due to either the diffusion of the fluorophores in and out from the part of the specimen from which the fluorescence is collected and conveyed to the fluorescence detector, the so-called observation value, or to physical/chemical reactions inducing changes in the emission rate, their analysis can be applied both to characterize the diffusion kinetics and to study the physical/chemical reactions affecting emission. Such techniques are collectively called fluorescence fluctuation spectroscopy (FFS) techniques. This sort of analysis is particularly useful in probing biological systems both in vitro and in vivo. Indeed, many biomolecules (e.g. enzymes) and substances (such as minerals or ions) that are present in traces in cells have a determinant role in the cellular metabolism, and techniques allowing to study their action at physiological concentrations are most desirable. The most popular FFS technique is fluorescence correlation spectroscopy (FCS), which was developed in the early 1970's (Gosch et al., 2005; Schwille, 2001), and consists in the analysis of the temporal correlations of fluorescence fluctuations in small molecular ensembles, combining maximum sensitivity with high statistical confidence. Among a multitude of physical parameters that are in principle accessible by FCS, it most conveniently allows to determine characteristic rate constants of fast reversible and slow irreversible reactions of fluorescently labelled biomolecules at very low (nanomolar and sub-nanomolar) concentrations. Moreover, the technique is a non-invasive one, which means that the above parameters can be assessed without perturbing the system equilibrium and, possibly, under physiological conditions. The technique also allows the determination of the mobility coefficients and local concentrations of fluorescent (or fluorescently labelled) molecules in their natural environments, provided that the exact value of the observation volume is known. More recently, at the end of the 1990's, another FFS technique, called photon counting histogram (PCH), was developed by Chen et al. (Chen et al., 1999; 2005). This technique is based on the analysis of the statistical distribution of the fluctuations in amplitude of the fluorescence rate and is thus suited to yield steady state parameters of the system to be investigated, such as the number of different fluorescent species present in a solution, the average number of molecules of each species in the observation volume, and their molecular brightness, rather than dynamic parameters. This technique is thus to a certain extent complementary to FCS. Until now, both FCS and PCH measurements have been made by using APDs or other single-photon counters as the light detectors. Information on either the time-correlations or the statistical distribution of the fluorescence intensity fluctuations have been derived by counting the detection events in consecutive and equally long time intervals. As single-photon counters must be used only in the single photon regime (that is keeping the overall count rate very low in order to be sure that impingement on the detector sensitive area of photons during their dead time is a negligibly rare event), this sampling procedure is extremely time consuming and the sampling interval width must not be too small. Namely, in the most frequent case of pulsed excitation, it has to correspond to tens or even hundreds of pulse periods, which means tens of microseconds or more. Obviously, this puts an inferior limit on the sensibility to fluorescence fluctuations occurring on time scales comparable, or shorter than, the sampling time. Conversely, many biological reactions occur on time scales of hundreds of nanoseconds to few microseconds (Nettels et al., 2007), that is slightly beyond the detection limit with the current procedures. We believe that the usage, for FFS techniques, of detectors able to count how many photons are contained in each fluorescence pulse could allow avoiding sampling. This will significantly and inherently enhance the temporal resolution of both techniques, virtually pulling it on the scale of the excitation source pulse period. Another advantage connected to the release of the single-photon regime operation condition is the possibility of performing statistically reliable measurements in much shorter times, which is notably beneficial for in-vivo analyses, as the samples quickly deteriorate. Recently, we have addressed the task of performing PCH analysis with SiPM detectors. Firstly, the standard PCH model developed for fitting data acquired with APDs (Chen et al., 1999; 2005) is not suitable to properly analyze the SiPM output; thus, the phenomenological model for describing the SiPM response presented above has been used to modify the PCH equations so as to take into account the deviations for ideal detection caused by the presence of optical cros-talk and sizeable DCR, leading to the following fitting model:

$$\Pi^\star \left(k; \bar{N}, \epsilon, x_t, \langle n \rangle_{dcr}\right) =$$

$$= \sum_{N=0}^{\infty} p_{\#}(N; \bar{N}) \sum_{m=0}^{k} C_{k,m}(x_t) \sum_{n=0}^{\infty} P_n(\langle n \rangle_{dcr}) p^{(N)}(m - n; \epsilon) \tag{27}$$

where x_t is the parameter representing the probability for a cell of triggering a neighbour cell by cross-talk and $C_{k,m}(x_t)$ is the Bernoullian-like distribution defined in Section 3.1 describing the probability of measuring k G-M avalanches out of n previously triggered, with $k > m$. Secondly, a two-photon excitation setup has been commissioned and tested by performing standard PCH measurements on Rodamine B water solutions with a single-photon avalanche diode (PDM50, MPD) and a PC board allowing for real-time PCH reconstruction (SPC150, Becker & Hickl GmbH). The setup relied on fluorescence excitation by means of a continuous-wave SESAM mode-locked Ti:sapphire laser (Tiger-ps SHG, Time Bandwidth Product) delivering 3.9 ps pulses of 840 nm wavelength at a repetition rate of 48 Mhz. A 1.25 NA, 100X oil-immersion microscope objective (Nikon) was used both to focus the excitation beam onto the sample and to collect the emitted light in epifluorescence configuration. The fluorescence signal was separated from the back-scattered excitation light by using a dichroic mirror. Residual stray light was removed by using a short-wavelength-pass filter with cut at 700 nm. The fluorescence was delivered to the detector, through a 1 mm diameter multimode fiber. With this setup, appreciably superPoissonian PCH distributions were obtained for <300 nM dye solutions. Finally, a series of feasibility studies on the same dye were performed by using as the detector a Hamamatsu SiPM (MPPC S10362-11-100C). Given the 48 MHz repetition rate of the laser light source, the SiPM is expected to detect a fluorescence light pulse every 20 ns. In such experimental conditions, the usual detector signal processing seemed unfeasible with the instrumentation at hand by both the integration method and the free-time signal acquisition method. Indeed, in the first case the CAEN V792N QDC used for the quantum-light statistics measurements presented above requires an integration gate starting at least 15 ns before the rise of the signal, meaning that the board would not be able to process every pulse; moreover, with a detector recovery time of $\sim$ 60 ns, the output signal may exhibit pile-up features, which in turn would lead to difficulties in reconstructing the photon statistics. In the second case, the DT5720A Desktop Digitizer at our disposal samples the output signal every 4 ns, which means that the signal integration would be performed using only 5 points. For these reasons, a completely different approach to process the detector signal was followed. Namely, the coincidence frequency $v(x)$ between a threshold scan of the SiPM output and a trigger signal v_T synchronous with the laser pulses (but with a threefold reduced frequency) has been acquired. The measured frequency can be described by the following relation:

$$v(x) = v_T \left(1 - e^{-t_C v_S \int_x^{+\infty} dy P(y)} \right) + v_T t_C v_S \int_x^{+\infty} dy P(y); \tag{28}$$

where $P(y)$ is the probability of measuring a value y of the detector output signal, $v(x)$ is the threshold value, v_S is the laser repetition rate and t_C is the coincidence time window. Equation 28 is based on the fact that a coincidence occurs both if after a trigger signal at least a SiPM signal over the threshold x is triggered within the coincidence time window or if the trigger signal rises within the coincidence time after the SiPM signal crosses the threshold. For small expectation values, Equation 28 can be approximated to

$$v(x) \approx 2 v_T v_S t_C \int_x^{+\infty} dy \, P(y). \tag{29}$$

Because of the Fundamental Theorem of Integration the detector output signal probability distribution can be written as

$$P(x) = \frac{-1}{2 v_T v_S t_C} \frac{dv(x)}{dx} \tag{30}$$

which can be processed in the same way as the experimental detector output signal histograms experimentally obtained in the case of quantum-light statistics measurements to yield the probability distribution of the number of detected photons (see Section 3). In this set of measurements, the detector was controlled by the CAEN SP5600 General Purpose Power Supply and Amplification Unit module with an integrated threshold discriminator. The digital NIM output of the discriminator was put in coincidence with a synchronism signal triggered by the laser by using an EG&G CO4010 module. The coincidence frequency was sampled by a CAEN VME V1718. The fluorescence emission of two solutions of rhodamine B in water, with concentrations of 2 μM and 200 nM, respectively, have been analyzed with the threshold scan method described above. For the 2 μM concentrated solution a 1 optical density neutral filter was placed in front of the fiber, to reduce the mean detected fluorescence intensity to the value of the 200 nM concentrated solution. This makes the comparison between the two resulting distributions easier. Indeed, for equal expectation value, the 200 nM solution PCH should feature larger deviations from Poisson distribution, therefore showing a higher tail. Hereafter, the pertaining results are briefly summarized and their significance on the way towards implementation of PCH with SiPMs is outlined. Using Equation 30 the multi-peak spectra reported in Fig. 14 were obtained from the staircases for the 200 nM and the 2 μM solution, respectively. The PCH were derived from the spectra by applying the same procedure implemented in the quantum-light statistics data analysis. The zero-photons frequency was determined by imposing the normalization condition:

$$\Pi^{\star}(0) = 1 - \sum_{k=1}^{L} \Pi^{\star}(k) \tag{31}$$

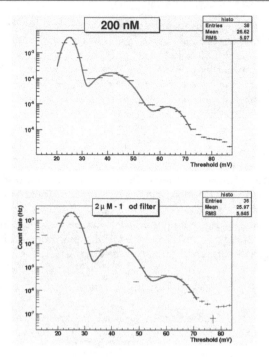

Fig. 14. Multipeak spectra of the fluorescence obtained from the threshold scans using Equation 30; in the upper panel spectrum from the 200 nM solution is shown, while in the lower panel the one obtained from the 2 μM solution is presented. Since the staircases has been acquired with a 2 mV step, histograms have 2 mV large bins. Spectra have been fitted with a multi-Gaussian function and the areas of each peak have been used as the PCH entries.

where L is the last detectable peak. The PCH are displayed in Figs. 15, for the 200 nM and the 2 μM solution, respectively. Fitting to Equation 27, with the average number of dark counts fixed to the experimental value preliminarily determined by performing a coincidence measurement of the dark counts, yields the fitting curves reported as dashed lines in Figs. 15.

The fitting parameters are summarized in Table 4.

Parameters	Concentrations (nM)	
	200	2000
$\bar{N}$	5.24 ± 0.03	71.70 ± 0.03
ϵ	$3.99 \times 10^{-3} \pm 2.4 \times 10^{-5}$	$1.536 \times 10^{-4} \pm 7 \times 10^{-7}$
x_t	$0.08085 \pm 5 \times 10^{-5}$	$0.0637 \pm 1 \times 10^{-4}$
$\langle n \rangle_{dcr}$	3.24×10^{-3}	3.24×10^{-3}

Table 4. Results for the fit of the PCH distributions. The errors listed are obtained from the fit procedure. The results for the mean number of particles and the molecular brightness are consistent with what expected, even though there is not exactly a factor 10 between the parameters of the two data sets.

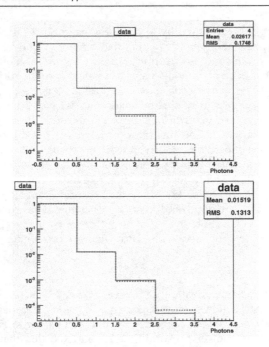

Fig. 15. Fit (red dashed line) of the experimental PCH data (blue continuous line) for 200 nM concentration (upper panel) and 2 μM concentration (lower panel).

The brightness and average particle number values are fairly consistent with the experimental conditions. Namely, the mean numbers of particles in the excitation volume differ by approximately one order of magnitude, reflecting the 10:1 concentration ratio between the two samples. To this purpose it should be noted that, while the concentration of the 2 μM solution could have been determined spectrophotometrically, 200 nM represents only a nominal value of the concentration of the less concentrated sample, obtained by dilution of the first, as the absorbance of it is below the sensitivity of our spectrophotometer (≈ 0.001) in this case. Also the difference in molecular brightness is roughly consistent with the insertion of the 1-optical-density filter. Indeed, the fact that the molecular brightness measured for the 2 μM concentrated sample is found to be more than twenty times lower could be ascribed to a slight misalignment of either the laser or the microscope objective during the measurement session. The latter lasted for many hours due to the necessity of determining with high statistical reliability the very low frequencies of coincidences occurring at a high threshold value. Even a slight degeneration in the quality of the optical setup alignment would determine a notable decrease in the two-photon absorption cross section, which scales as the square power of the excitation intensity. This would be reflected in the brightness value. A broader focus is also consistent with the mean number of molecules in the excitation volume detected for the 2 μM solution, which is slightly larger than expected. Finally, the difference between the cross-talk values is consistent with a lower over-biasing condition for the 2 μM set, suggesting that a worsening of the photon detection efficiency could also have contributed to the lower molecular brightness. It must be admitted that the above presented procedure is very far from ideal pulse-by-pulse PCH acquisition. However, these preliminary data demonstrate that despite the relatively high dark count rates with respect to APDs and the additional artifice constituted by optical cross-talk, SiPMs are detectors capable of discriminating the tiny deviations from Poisson distribution typically displayed by the PCH of diffusing fluorophores in solution at concentrations relevant for biophysical applications. Moreover, we notice that our fitting model yields consistent values of brightness and average fluorophore number. The above conclusions suggest that pulse-by-pulse reconstruction of PCH should be straightforwardly obtained with our detector and electronics apparatus for excitation rates <10 MHz, according to both the recovery time of the SiPM detector and the requirements of the CAEN integrator. An optical pulse-picker may be used to customize the Ti:sapphire repetition rate. Even working at such rates would allow to speed up PCH reconstruction by more than two orders of magnitude with respect to state-of-art APD-based methods. Alternatively, high-repetition rate measurements might be performed by reconstruction methods based on the analysis of the detector free-run output signal, as recorded by fast-sampling-rate digitizers. In such a case, the issues connected with detector pile-up effects would be overcome, thus confining the limits on the acquisition rate to the excitation rate and to the intrinsic decay time of the fluorophores under analysis.

5. Conclusions

We have presented the characterization of a class of photodetectors, namely the Silicon Photomultipliers, that has acquired great importance in the last decades because of its good photon-counting capability. Besides this feature, the compactness, robustness, low cost, low operating voltage and power consumption are added value that would make these detectors suitable for many applications. To this aim, we have implemented an exhaustive characterization protocol providing a quantitative evaluation of the main figures of merit these detectors are endowed with. In particular, we have studied the dependence of the photon detection efficiency, gain, dark count rate and optical cross-talk on temperature. The protocol has been applied to three SiPM detectors produced by different manufactures and the results have been then employed in a partnership with CAEN to develop the SP5600 General Purpose Power Supply and Amplification Unit module.

The versatility of SiPMs has been investigated in the reconstruction of the statistics of light states: two complementary models that take into account the DCR and the cross-talk effects have been developed in order to describe the output signal of the detectors. The correctness of these methods has been experimentally validated by illuminating the SiPM sensor with a single-mode pseudo-thermal light. However, further improvements could be achieved by including in the models the afterpulsing effect, which would become important in high-rate events.

The applicability of SiPMs in Fluorescence Fluctuation Spectroscopy studies has been preliminary evaluated by analyzing Rhodamine solutions by means of a Two-Photon Excitation setup: the capability of the sensors to reconstruct Photon Counting Histogram and in particular to provide the estimation of the mean number of fluorophores in the excitation volume and their molecular brightness has been proved.

6. References

Afek, I.; Natan, A.; Ambar, O.; & Silberberg, Y. (2009). Quantum state measurements using multipixel photon detectors. *Phys. Rev. A.*, Vol. 79, No. 4, April 2009, 043830, ISSN 1050-2947.

Agliati, A.; Bondani, M.; Andreoni, A.; De Cillis, G.; & Paris, M. G. A. (2005). Quantum and classical correlations of intense beams of light investigated via joint photodetection. *J. Opt. B Quantum Semiclass. Opt.*, Vol. 7, No. 12, November 2005, S652–S663, ISSN 1464-4266.

Akindinov, A. V.; Martemianov, A. N.; Polozov, P. A.; Golovin, V. M.; & Grigoriev, E. A. (1997). New results on MRS APDs. *Nuclear Instruments and Methods in Physics Research Section A: Accelerators, Spectrometers, Detectors and Associated Equipment*, Vol. 387, No. 1-2, March 1997, 231–234, ISSN 0168-9002.

Akindinov, A; Bondarenko, G.; Golovin, V.; Grigoriev, E.; Grishuk, Yu.; Malkevich, D.; Martemiyanov, A.; Ryabinin, M.; Smirnitskiy, A.; & Voloshin, K. (2005). Scintillation counter with MRS APD light readout. *Nuclear Instruments and Methods in Physics Research Section A: Accelerators, Spectrometers, Detectors and Associated Equipment*, Vol. 539, No. 1-2, February 2005, 172–176, ISSN 0168-9002.

Andreoni, A. & Bondani, M. (2009). Photon statistics in the macroscopic realm measured without photon counters. *Phys. Rev. A*, Vol. 80, No. 1, July 2009, 013819, ISSN 1050-2947.

Arecchi, F. T. (1965). Measurement of the Statistical Distribution of Gaussian and Laser Sources. *Phys. Rev. Lett.*, Vol. 15, No. 24, December 1965, 912–916, ISSN 0031-9007.

Bondani, M.; Allevi, A.; & Andreoni, A. (2009). Light Statistics by Non-Calibrated Linear Photodetectors. *Adv. Sci. Lett.*, Vol. 2, No. 4, December 2009, 463–468, ISSN 1936-6612.

Bondani, M.; Allevi, A.; Agliati, A.; & Andreoni, A. (2009). Self-consistent characterization of light statistics. *J. Mod. Opt.*, Vol. 56, No. 2, January 2009, 226–231, ISSN 0950-0340 print/ 1362-3044 online.

Chen, Y.; Müller, J. D.; So, P. T. C.; & Gratton, E. (1999). The Photon Counting Histogram in Fluorescence Fluctuation Spectroscopy. *Biophys. J.*, Vol. 77, No. 1, July 1999, 553–567, ISSN 0006-3495.

Chen, Y.; Tekmen, M.; Hillesheim, L.; Skinner, J.; Wu, B., & Muller, J. D. (2005). Dual-color photon counting histogram. *Biophys. J.*, Vol. 88, No. 3, March 2005, 2177–2192, ISSN 0006-3495.

Cova, S; Ghioni, M.; Lacaita, A.; Samori, C.; & Zappa, F. (1996). Avalanche photodiodes and quenching circuits for single-photon detection. *Appl. Opt.*, Vol. 35, No. 12, April 1996, 1956–1976, ISSN 1559-128X print/ 2155-3165 online.

Du Y. & Reti'ere, F. (2008). After-pulsing and cross-talk in multi-pixel photon counters. *Nuclear Instruments and Methods in Physics Research Section A: Accelerators, Spectrometers, Detectors and Associated Equipment*, Vol. 596, No. 3, November 2008, 396–401, ISSN 0168-9002.

Eckert, P.; Schultz-Coulon, H.-C.; Shen, W.; Stamen, R.; & Tadday, A. (2010). Characterisation studies of silicon photomultipliers. *Nuclear Instruments and Methods in Physics Research Section A: Accelerators, Spectrometers, Detectors and Associated Equipment*, Vol. 620, No. 2-3, August 2010, 217–226, ISSN 0168-9002.

Goetzberger, A.; McDonald, B.; Haitz, R. H.; & Scarlett, R. M. (1963). Avalanche Effects in Silicon p-n Junctions. II. Structurally Perfect Junctions. *Journal of Applied Physics*, Vol. 34, No. 6, June 1963, 1591–1600, ISSN 0021-8979.

Gösch, M. & Rigler, R. (2005). Fluorescence correlation spectroscopy of molecular motions and kinetics. *Adv. Drug. Del. Rev.*, Vol. 57, No. 1, January 2005, 169–190, ISSN 0169-409X.

Lutz, G. (1995). Silicon radiation detectors. *Nuclear Instruments and Methods in Physics Research Section A: Accelerators, Spectrometers, Detectors and Associated Equipment*, Vol. 367, No. 1-3, December 1995, 21–33, ISSN 0168-9002.

Mandel, L. & Wolf, E. (1995). *Optical Coherence and Quantum Optics*, Cambrige University Press, ISBN 0521417112, New York.

McKay, K. G. (1954). Avalanche Breakdown in Silicon. *Phys. Rev.*, Vol. 94, No. 4, May 1954, 877-884.

Nettels, D.; Gopich, I. V.; Hoffmann, A.; & Schuler, B. (2007). Ultrafast dynamics of protein collapse from single-molecule photon statistics. *Proc. Nat. Acad. Sci. USA*, Vol. 104, No. 8, February 2007, 2655–2660, ISSN 0027-8424.

Oldham, W. G.; Samuelson, R. R.; & Antognetti, P. (1972). Triggering phenomena in avalanche diodes. *IEEE Trans. Electron. Dev.*, Vol. 19, No. 9, September 1972, 1056-1060, ISSN 0018-9383.

Ramilli, M.; Allevi, A.; Chmill, V.; Bondani, M.; Caccia, M.; & Andreoni, A. (2010). Photon-number statistics with Silicon photomultipliers. *J. Opt. Soc. Am. B*, Vol. 27, No. 5, May 2010, 852–862, ISSN 0740-3224.

Schwille, P. (2001). Fluorescence correlation spectroscopy and its potential for intracellular applications. *Cell Biochem. Biophys.*, Vol. 34, No. 3, June 2001, 383–408, ISSN 1085-9195.

Sciacca, G.; Condorelli, E.; Aurite, S.; Lombardo, S.; Mazzillo, M.; Sanfilippo, D.; Fallica, G.; & Rimini, E. (2008). Crosstalk characterization in Geiger-mode avalanche photodiode arrays. *IEEE Electron. Dev. Lett.*, Vol. 29, No. 3, March 2008, 218-220, ISSN 0741-3106.

Tarolli, A.; Dalla Betta, G.-F.; Melchiorri, M.; Piazza, A.; Pancheri, L.; Piemonte, C.; & Zorzi, N. (2010). Characterization of FBK SiPMs under illumination with very fast light pulses. *Nuclear Instruments and Methods in Physics Research Section A: Accelerators, Spectrometers, Detectors and Associated Equipment*, Vol. 617, No. 1-3, May 2010, 430–431, ISSN 0168-9002.

Van Der Marel, R. P. & Franx, M. (1993). A new method for the identification of non-gaussian line profiles in elliptical galaxies. *Astrophysical Journal*, Vol. 407, No. 2, April 1993, 525–539, ISSN 0004-637X.

Zambra, G.; Bondani, M.; Spinelli, A. S.; Paleari, F.; & Andreoni, A. (2004). Counting photoelectrons in the response of a photomultiplier tube to single picosecond light pulses. *Rev. Sci. Instrum.*, Vol. 75, No. 8, August 2004, 2762–2765, ISSN 0034-6748 print/1089-7623 online.

7

Far-Infrared Single-Photon Detectors Fabricated in Double-Quantum-Well Structures

Takeji Ueda and Susumu Komiyama
The University of Tokyo
Japan

1. Introduction

Sensitive infrared (IR) detectors are key elements for both fundamental research and applications, and development of sensitive IR detectors is one of intensive research subjects in resent years (Rogalski, 2002; Tidrow, 2000). The activities are motivated by both basic and applied research fields (Prochazka, 2005; Rogalski, 2002). For certain applications, an ultimately high sensitivity reaching a photon-counting level is indispensable; for instance, passive high-resolution microscopy, catching IR photons spontaneously emitted by a living cell or a small number of biomolecules, may be accessible only with such ultimately sensitive detectors.

Photon-counters are routinely used in visible and near infrared regions (Prochazka, 2005). In the long-wavelength range ($\lambda > 10 \ \mu$m, frequencies lower than 30 THz), called terahertz (THz) region, however, photon energies are far smaller ($h\nu < 124$ meV for $\lambda > 10 \ \mu$m) and the single-photon detection is no longer trivial. Nevertheless, the THz region is one of the richest areas of spectroscopy of matters, encompassing the rotational spectra of molecules, vibrational spectra of molecules, liquids and solids, and the electron energy spectra in semiconductor nanostructures and superconducting energy gap in metals. Hence, sensitive microscopy of matters with high spatial resolution in this spectral region would provide a unique powerful tool for investigation of matters. Sensitive observation is also strongly demanded in astrophysics (Nakagawa et al, 2007).

In the last decade, a variety of novel detection schemes have been proposed (Astaviev et al., 2002; Day et al., 2003; Hashiba et al., 2006; Komiyama et al., 2000; Schoelkopf et al., 1999; Wei et al., 2008). Among them, only semiconductor quantum devices have demonstrated single-photon detection (Astaviev et al., 2002; Hashiba et al., 2006; Komiyama et al., 2000): The experimentally achieved noise equivalent power (NEP), less than 1×10^{-19} W/Hz$^{1/2}$, is by several orders of magnitude lower than those of any other detectors.

These Semiconductor photon-counters exploit a novel scheme, in which the photon absorption event leads to generation of a long-lived unit charge that is probed by a single-electron transistor (SET). The use of SETs, however, might restrict their application because the operation is limited to ultra-low temperatures (<1K). From the viewpoint of broader application, therefore, photon-counters usable at elevated temperatures are highly desirable. From the viewpoint of importance of the spectra of matters, it is also extremely important to fill the gap region between the near infrared and the far infrared regions; that is, the long-wavelength infrared (LWIR) and the mid-infrared (MIR) regions.

In this chapter, another type of charge-sensitive infrared phototransistors in 5-50 μm wavelength range is described. The detectors are called charge-sensitive infrared phototransistors (CSIPs). The detectors can have ultra-sensitivity reaching singe-photon detection level, as well as ultra-broad dynamic range (> 10^6, from attowatts to beyond picowatts)(Ueda et al., 2008). CSIPs can be operated reasonable temperatures (~25 K at 15 μm, depending on wavelength) (Ueda et al., 2009). The excellent noise equivalent power (NEP=6.8x10^{-19} W/Hz$^{1/2}$) and specific detectivity (D^*=1.2x10^{15} cmHz$^{1/2}$/W) are demonstrated for $λ$=14.7 μm, which are by a few orders of magnitude superior to those of the other state-of-the-art detectors (Ueda et al., 2008). In addition, the simple planar structure is, similarly to CMOS sensors, feasible for array fabrication and will even make it possible to monolithically integrate with reading circuit.

2. Detection scheme and device structure

In CSIPs, an electrically isolated island of a QW is photoexcited to serve as a gate to a remote two-dimensional electron gas (2DEG) conducting channel. As schematically shown in Fig. 1 (a), photoexcited electrons escape the isolated QW island leaving holes behind. The photo electrons are driven to the 2DEG conducting channel yielding photocurrents. Another effect larger than this direct photocurrent arises from the positive charge left on the QW island, which, through capacitive coupling, increases the electron density in the 2DEG channel and thus its conductance. The effect persists until the excited electrons recombine with holes in the isolated island, serving as an amplification mechanism. CSIPs are thus charge-sensitive phototransistor, in which a QW island works as a photosensitive floating gate.

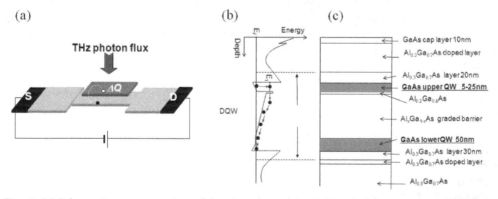

Fig. 1. (a) Schematic representation of the charge sensitive infrared phototransistor (CSIP) as a phpto-active field-effect transistor. (b) The energy diagram of double-quantum-well system. (c) Crystal strucuture for detection wavelength of 15 μm.

The upper QW is so designed that the energy spacing between the ground subband and the first excited subband is $ΔE$ =84 meV (for detection wavelength of $λ$=15 μm). When radiation with photon energy of $ΔE$ is incident on the isolated QW, electrons are excited to the first excited subband, where the thin tunnel barrier layer stands as schematically depicted in Fig. 1 (b). The electrons, having tunneled out of the QW, fall down the electrostatic potential slope in the graded barrier layer until they eventually reach the 2DEG channel to be absorbed there. This causes isolated QW island to be positively charged. Through capacitive coupling, the pile-up positive charge in the isolated island increases the electron density of the lower 2DEG channel leading to an increase in conductance.

The crystals are grown by molecular-beam epitaxy (MBE). The crystal structure is shown in Fig. 1 (c). Undoped GaAs (~400 nm) and 10 periods of an $Al_{0.3}Ga_{0.7}As$ (20 nm)/GaAs (2 nm) superlattice were first grown on the semi-insulating substrate. Then, a $1x10^{24}$ m^{-3} Si doped $Al_{0.3}Ga_{0.7}As$ layer (10 nm) and an $Al_{0.3}Ga_{0.7}As$ spacer layer (30 nm) were grown as a remote-doping layer to supply electrons to the successively grown undoped GaAs layer (20 nm), which serves as the lower QW in the DQW system. The 150-nm thick compositionally graded $Al_xGa_{1-x}As$ barrier (GB) and a 2-nm thick tunnel barrier (TB) are grown as the isolation barrier between the lower and upper QWs. Here, the growth was suspended between the GB and TB to change the Al beam. The GaAs layer as the upper QW was successively grown. Finally, an $Al_{0.3}Ga_{0.7}As$ spacer layer and an $Al_{0.3}Ga_{0.7}As$ uniform Si doping layer ($1x10^{24}$ m^{-3}, 40 nm) were grown to supply electrons to the upper QW. The surface is covered with a 10-nm-thick GaAs cap layer. The superlattice barrier below the lower QW improves the roughness of the GaAs/AlGaAs interface. The detection wavelength is controlled by changing the design width of upper QW (Ueda et al., 2011).

(a)   (b)

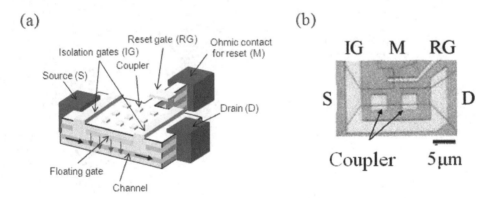

Fig. 2. (a) Schematic representation of QWs and ohmic contacts. The upper QW is electrically isolated by negative biasing of surface gates. (b) A microscope image of the device with a 16x4 μm^2 isolated island formed by the upper QW.

As shown in Fig. 2, the device consists of a wet-etched DQW mesa, alloyed AuGeNi ohmic contacts, Au/Ti Schottky gates (the isolation gate, IG, and reset gate, RG), and Au/Ti photo-coupler (antenna). The device is fabricated with standard electron-beam lithography technique. The 2DEG layer in the both of the QWs are normally connected by ohmic contacts, and can be electrically isolated by biasing metal isolation gates. The active areas of devices are freely defined in lithography. The coupler is used to cause intersubband transition by generating electric filed normal to the plane of the QW against the normally incident radiation. It should be mentioned that the simple planar structure of CSIPs is feasible for array fabrication (fabricated example is shown in Fig. 3). The FET structure would have an advantage of future monolithic integration with reading circuits.

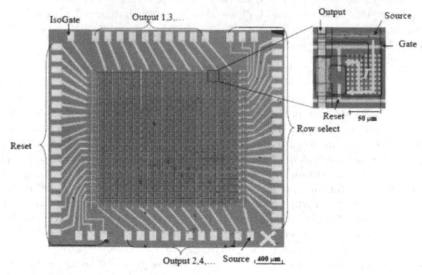

Fig. 3. 20x 20 CSIP focal plane array

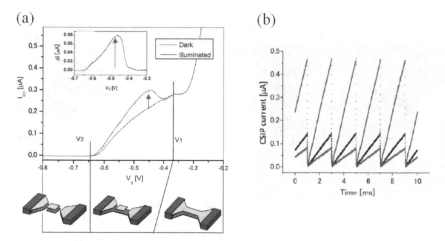

Fig. 4. (a) I-V measurement of a CSIP with scanning gates (IG and RG) bias. (b) Integration ramps of the measured photo-current against incident radiation with different intensities. Releasing the accumulated charge in the floating gate with reset gate pulses restore the detector to its original state.

In actual practice of operation, the optimized metal-gate bias is found in gate-bias-dependent IV curve where photosignal reaches maximum amplitude as shown in Fig. 4 (a). The signal appears when FET is formed: upper QW is electrically isolated and the accumulated photoholes induce larger current flowing in the lower QW.

As in Fig. 4 (b), to avoid the saturation of the signal, the accumulated charge is released to the reservoir by applying a brief positive pulse (e.g., +0.3V, 1 μs) to the RG, and the CSIP is reset to the original highly sensitive state (An et al., 2007). Conventional photoconductive detectors yield a certain amplitude of photocurrent that is proportional to the incident radiation intensity. The scheme of CSIPs is different: The source-drain current continues to increase with time under steady illumination, where the speed at which the current increases, $a=\Delta I/\Delta t$, is proportional to the incident radiation intensity.

This photoresponse can be interpreted by increase of electron density in lower QW induced by capacitively coupled photoholes stored in the isolated upper QW. The unit increment of current I_e induced by one photohole in the isolated upper QW (area of $L \times W$) is given by:

$$I_e = \frac{e\mu V_{SD}}{L^2} \tag{1}$$

where e is the unit charge, μ the electron mobility of lower QW, V_{SD} the souce-drain (SD) voltage, L the length of constricted channel. For example, unit increment I_e = 3 pA is given for μ = 1 m^2/Vs, V_{SD} = 10 mV and L = 16 μm. The signal I_e persists as long as a photohole stays in upper QW. By setting the lifetime τ = 1 s, the amplification factor, or photoconductive gain, is given as $G = \tau I_e/e = 1.8 \times 10^7$. This value is comparable to that of

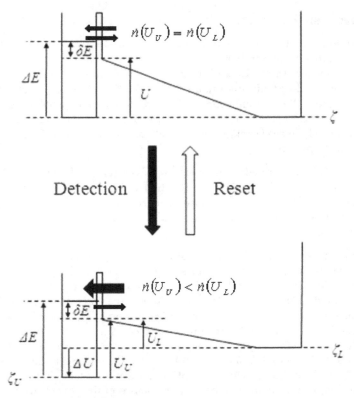

Fig. 5. Energy diagram illustrating photo-saturation.

photomultiplier tubes. The increase of the current is proportional to the number of accumulated photoholes p:

$$\Delta I = p I_e \tag{2}$$

Under steady illumination, p is a linearly increasing function of time if the lifetime of photoholes is longer than the relevant time of integration. The number of photoholes, of course, does not increase infinitely, but reach a saturated value. The saturation is caused by balance between generation and recombination speeds of photoholes which change with deformation of the potential profile due to accumulating positive charges in upper QW, as shown in Fig. 5. The potential drop is given by $\Delta U = pe^2 d / \varepsilon L W$, where $d = 150$ nm is the distance between upper and lower QWs, $\varepsilon = 12 \times (8.85 \times 10^{-12})$ F/m is the electric permittivity of GaAs. The number of photoholes is described by the rate equation:

$$\frac{dp}{dt} = \eta \Phi - \frac{p}{\tau} \tag{3}$$

where η is quantum efficiency, Φ is the incident photon flux (photons/s), and τ is the lifetime of photoholes. The first and second term in the right-hand side of Eq. 3 refer to the generation and recombination speed of photoholes, respectively.

3. All-cryogenic spectrometer

It is not trivial to accurately characterize ultrasensitive detectors in far infrared region. In usual setup, a monochromator system is placed at room temperature. Then, in case of ultrasensitive detection in far-IR and THz region, the measurements suffer from the thermal background blackbody radiation (BBR) at 300K that can be much stronger than the light guided from the external source. This caused a serious problem for the study of extremely sensitive CSIPs, the intrinsic characteristics of which may be significantly affected by relatively weak background radiation. It was also difficult to determine accurate excitation spectrum.

We characterize CSIPs in reliable manners by developing a home-made all-cryogenic spectrometer (4.2K) (Ueda, 2007). As illustrated in Fig. 6 (a), the all-cryogenic spectrometer consists of an emitter, a parabolic mirror, a rotating diffraction grating, and the detector. The whole system is assembled on a 1cm-thick copper base plate and inserted in a 60mmφ metal pipe.

The emitter is a 1kΩ thin metal-film chip resistor (Fig. 6 (b)), consisting of a 1x2mm² resistive metal film coated on a 0.5mm-thick alumina substrate and covered with a ~20μm-thick glass overcoat layer. The resistance (1kΩ) does not change appreciably in the range of temperature studied (4.2K up to 244K). The resistor is sealed in a small metal chamber (8 mm-length and 4 mm-diameter, not shown in Fig.6 (b)) and suspended in vacuum (~3x10⁻⁴ Pa) by current leads (0.2mmφ Manganin wires). We suppose that the resistor serves approximately as a black body when heated with current, since the emissivity of the overcoat glass layer is beyond 90% in the LWIR range. The emitted radiation is transmitted through a 625μm-thick GaAs window (optical transmittance of 50-60% in a 5-15 μm wavelength region) of the small chamber. The resistance of the current leads is ~30 Ω over the total length (2 m) in the cryostat. Hence the electrical power is substantially dissipated by the chip resistor. The emitter temperature, $T_{emitter}$, is monitored with a thermocouple, Chromel-Au/Fe (0.07 %) glued to the backside of the resistor (Fig.6 (b)). The values of $T_{emitter}$ will be shown as a function of the input electrical power P_{in} in the inset of Fig.9 (b).

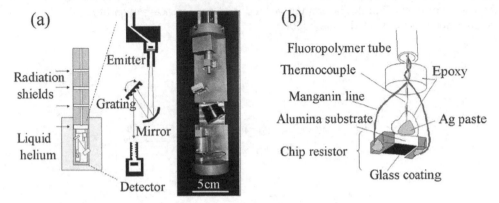

Fig. 6. (a) All cryogenic spectrometer; schematic representation and a photograph. (b) A 1kΩ thin metal film chip resistor serving as a black-body radiator. The emitter is packaged in a small vacuum metal chanber with a GaAs window.

The parabolic mirror collimates broadband BBR from the emitter and directs it to the diffraction grating, which will be described in the next paragraph. The angle of diffraction grating is controlled mechanically from the outside of the cryostat. The radiation diffracted at the grating is guided to the detector. The detector is installed in a small copper box with a 4 mmφ and 20 mm-long metal pipe at the entrance. The inner wall of the entrance pipe is screw patterned so as to prevent stray lights from entering the detector box.

A home-made diffraction grating is used. Commercially available diffraction gratings are usually composed of materials of different thermal expansion coefficients, like an epoxy substrate coated with aluminum films. They are hence unreliable for the use at cryogenic temperatures. We fabricated a diffraction grating using a GaAs crystal. A GaAs crystal surface was patterned into 7.5μm-wide and 7.5μm-deep trenches at 15 μm-period via mesa etching with a solution of $H_2PO_4 : H_2O_2 : H_2O = 4 : 1 : 40$ (etching rate: 40 nm/min) at 25 ºC. The etching was made for 4.5 hours. The slow rate and the long time of etching were chosen so as to round the grove edges through under-etching the resist mask. The rounding of grove edges was necessary for obtaining a reasonable amplitude for the 1st-order diffraction by reducing the 0th-order diffraction (specular reflection). Finally the trenched GaAs surface was coated with a 20nm-Ti/300nm-Au layer via vacuum evaporation.

The whole spectrometer system is immersed in liquid helium (T=4.2K). The 300K BBR from the warm part of the cryostat is blocked by placing copper plates at a 10 cm-interval inside the metal pipe. The emitter chamber is made vacuum by pumping through a thin, 4-mm diameter, metal pipe. The pipe is narrowed to 2mmφ at the entry to the chamber and bended at two places for minimizing the 300K-BBR entering through the vacuum pipe.

The wavelength resolution of the present optical setup was evaluated to be $\Delta\lambda_g$=1.0 μm (full width at half maximum, FWHM) by using a Fourier Transform Infrared Spectroscopy (FTIR) system.

4. Spectral response

The spectral response of the detector was studied by varying the angle of the diffraction grating of All cyogenic spectrometer at a step of $\Delta\theta = 0.4$ degree (corresponding change in the wavelength of 0.2 µm). The emitter was fed with an electrical power of P_{in}=0.1 W, where the thermocouple indicated $T_{emitter}$=130K.

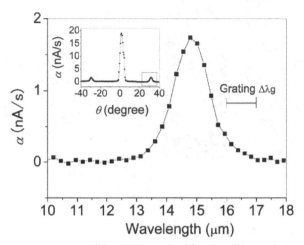

Fig. 7. The detection spectrogram of the CSIP studied by the all cryogenic spectrometer, where the instrumental resolution is $\Delta\lambda_g$ = 1µm as marked by the bar. The inset displays the original spectral chart.

The inset of Fig. 7 shows the current increases $a=\Delta I/\Delta t$ against the angle of grating θ, where the central peak (θ=0) corresponds to the 0th order diffraction and the two symmetrically located side peaks are the $\pm1^{st}$ order diffractions. Figure 7 elucidates the spectrogram of the right-hand side peak, where the horizontal axis is scaled by the wavelength. The peak wavelength is λ=14.7 µm, and the apparent spectral bandwidth (FWHM) is $\Delta\lambda_{app}$= 1.6 µm. By noting the instrumental resolution of $\Delta\lambda_g$=1.0 µm, we evaluate the true bandwidth to be $\Delta\lambda_d$ ~1.0 µm (Ueda et al, 2007).

Similarly, spectral responses for CSIPs with different upper QW widths, i.e. subband energy, have been studied (Ueda et al., 2011). As shown in Fig. 8, CSIPs cover the wavelength range of 10-50 µm, where there was no available ultrasensitive detector, reaching single-photon counting level.

5. Single-photon signal and dynamic range of CSIPs

Because the incident radiation intensity to the detector is not accurately determined by the experimental setup shown in Fig. 6 (a), the detector box was placed immediately in front of the emitter chamber as illustrated in the inset of Fig. 9 (a). Knowing the geometry, we can derive absolute values of the incident radiation intensity as will be discussed later. (Though wavelength selection is not made, only a narrow detection band (Fig. 7) contributes to the photo-response.)

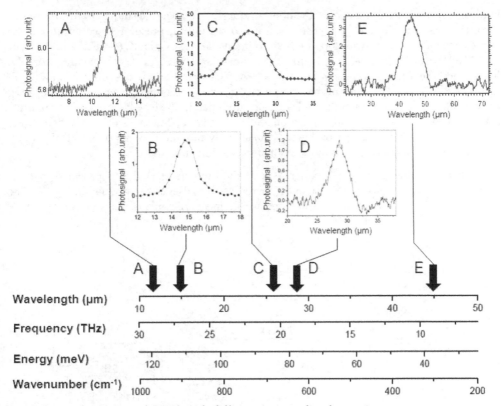

Fig. 8. Spectral response of CSIPs with different intersuuband energies.

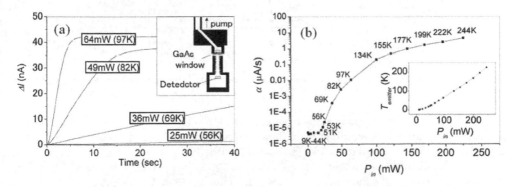

Fig. 9. (a) Time traces of the photo current, ΔI, obtained in different incident-radiation intensities, where the emitter directly faces the detector as shown in the inset. The electric power fed to the emitter, P_{in}, and the emitter temperature, $T_{emitter}$, are indicated for the respective curves. (b) The rate of photo-current increase ($a=\Delta I/\Delta t$) vs. the input electric power fed to the emitter (P_{in}). The emitter temperature $T_{emitter}$ is also indicated. The inset shows $T_{emitter}$ as a function of P_{in}, studied by the thermocouple attached to the emitter.

Figure 9 (a) displays time traces of the photo-induced change in the source-drain current, ΔI, obtained after the detector has been reset at t=0 sec with a pulse of 1μsec duration. The curves are taken with four different radiation intensities ($T_{emitter}$=56K, 69K 82K and 97K) for P_{in}= 25mW~64mW. It is clearly seen that the increase of current levels off deviating from the linear increase when ΔI exceeds around 20 nA. The current increases linearly in a small range of ΔI <20 nA. The slope of the linear increase, $a=\Delta I/\Delta t$, is the signal intensity indicating the rate at which photo-generated holes are accumulated in the isolated upper QW. Figure 9 (b) shows $a=\Delta I/\Delta t$ as a function of the input electric power P_{in} fed to the emitter, where $T_{emitter}$ is marked on each data point. Bachground blackbody radiation corresponding to 50 K still remains, but fortunately, the analysis is not afected by the radiation.

Though not visible in Fig.9 (a), closer look at the time traces makes it probable that the current does not increase smoothly but it increases stepwise, as exemplified in Fig. 10. The data of Fig.10 are taken with a time constant of measurements of Δt=1sec for P_{in}=1mW ($T_{emitter}$=14K).

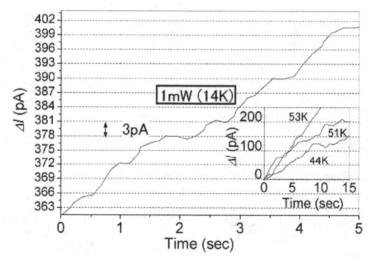

Fig. 10. Time trace of _I obtained with Temitter=14 K, which is represented in a magnified scale.

Figure 11 shows the histogram of the frequency of step-wise change ($\Delta I_{plateau}$) between clear plateaus (omitting small winding in continuous slope), occurring in the curves obtained at different low illumination intensities ($T_{emitter}$<50K). It is likely that the histogram consist of the Gaussian distributions with the mean values of $<\Delta I>$=μ= 3pA, 6pA, 9pA and 12pA. It is hence suggested that the increase of ΔI takes place with a step of ΔI_{Step} =3pA.

In the present scheme of photo detection (Figs. 1 and 2), the current through the lower QW layer is expected to increase by Eq. (1). The values of the mobility of the lower QW layer μ = 4.3x10³cm²/V, source-drain bias voltage V_{SD}=10mV, channel length L=16μm yield ΔI_e =2.7pA, which is close to the experimentally observed amplitude of the unit step ΔI_{Step} =3pA in the above.We note in Fig. 9 (a) that the detector exhibits linear response for the excited holes less than 20 nA/3 pA = 7x10³.

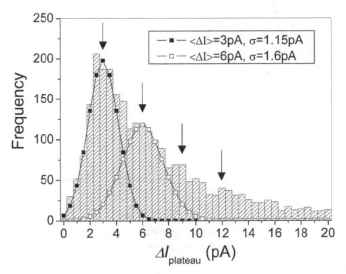

Fig. 11. The histogram of the stepwise change of current ($\Delta I_{plateau}$), occurring in ΔI versus t curves. Fitting lines represent the Gaussian distribution with $<\Delta I>$ the mean value and σ the standard deviation.

The quantum efficiency, $\eta = \{a/\Delta I_{Step}\}/\Phi$, is an important parameter defining the ratio of the rate of detected photons, $a/\Delta I_{Step}$, to that of the total incident photons, Φ. Here, the total incident photon flux is given by

$$\Phi = (\varepsilon f_{att}) \left[\frac{B(\lambda,T)\Omega \Delta \lambda_d}{hc/\lambda} \right] S_{em} \qquad (4)$$

where $B(\lambda,T) = (2hc^2/\lambda^5)\{\exp(hc/\lambda k_B T)-1\}^{-1}$ is the Plank's formula with h the Planck constant, k_B the Boltzmann constant and c the light velocity, $\Lambda\lambda_d=1\mu m$ is the detector bandwidth and $\Omega=S_{iso}/d^2 = 1.2\times10^{-7}sr$ is the solid angle with the active detector area of $S_{iso}=WL_{iso}=64\mu m^2$ with an emitter/detector distance of $d=2.3cm$. Here, $S_{em}=8mm^2$ is the effective emitter area, ε ~1 is the emissivity of the emitter (glass surface) , $f_{att}=0.45$ is the attenuation due to reflection at the GaAs window and $hc/\lambda = h\nu = 83$ meV is the photon energy ($\lambda=14.7\mu m$). Knowing these parameter values, we can derive the photon flux, Φ, from $T=T_{emitter}$.

The curve of a versus P_{in} ($T_{emitter}$) in Fig. 9 (b) is re-plotted in terms of the count rate, $a/\Delta I_{Step}$, and the incident photon flux, Φ, in Fig.12. The linear relationship between $a/\Delta I_{Step}$ and Φ assures the validity of our analysis, and indicates an efficiency of $\eta =2\pm 0.5$ %. The dynamic range of linear response is ~10^7 ($\Phi=4\times10/sec$~$1\times10^8/sec$, or $P=\Phi h\nu=5.3\times10^{-19}W$ ~ $1.3\times10^{-12}W$). Note that the maximum radiation power in Fig. 6 ($\Phi = 1\times10^8/sec$ or $P=\Phi h\nu=1.3\times10^{-12}W$) is not determined by the capability of the detector but is restricted by the emitter ($T_{emitter}$ =244K with P_{in}=225mW). As discussed later, the detector is expected to show linear response up to much higher power levels ($P=10^{-6}W$) in an appropriate condition of reset pulses, suggesting a much wider dynamic range reaching 10^{13}.

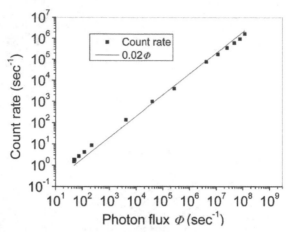

Fig. 12. Count rate of the photosignal vs. incident photon flux Φ. Photon fluxes of $\Phi=10^2$ photons/sec corresponds to $P_{in} = \Phi h\nu = 1.3\times10^{-18}$W.

6. Figures of merit

The current responsivity of the detector, $R=\Delta I/(h\nu\Phi)$, is the photocurrent amplitude ΔI divided by the incident radiation power. In the CSIP, $\Delta I=a\Delta t_I$ is proportional to the integration time, Δt_I, which can be arbitrarily chosen. If we take a typical reset interval, in a range of Δt_I=10msec~1sec, we derive $R=\eta(\Delta I_{Step}/h\nu)\Delta t_I = (4\times10^4$~$4\times10^6)$ A/W by noting $\eta=\{a/\Delta I_{Step}\}/\Phi$ with ΔI_{Step} =3pA and η=0.02. The responsivity is thus by many orders of magnitude larger than that of well known QWIPs, for which R is typically ~0.5A/W (Levine, 1993; Yao, 2000). We can also write the responsivity as $R=\eta(e/h\nu)g$ by using the "photoconductive gain" g, which refers to the effective number of electrons generated by a single photon. We mention that the large value of R comes from the extremely high photoconductive gain, $g=(\Delta I_{Step}/e)\Delta t_I = 1\times10^5$~$1\times10^7$, in the CSIP.

Noise equivalent power (NEP) is defined by

$$NEP = \frac{P}{SNR\sqrt{\Delta f}} \tag{5}$$

where P is the power of incident radiation, SNR is the signal to noise ratio, and $\Delta f=1/(2\Delta t)$ is the frequency bandwidth of measurements with Δt the averaging time. We evaluate Eq.(5) by taking three curves of ΔI versus t shown in the inset of Fig.7 as an example. These curves are taken for $T_{emitter}$=44K (P_{in}= 2.8aW), 51K (4.5aW) and 55K (5.0aW). Considering the time constant τ =1sec of measurements, we evaluate the noise from $\sqrt{\langle(\alpha-\bar{\alpha})^2\rangle}$, where $\bar{\alpha}$ is the average slope (count rate) of each curve for a long time (>15 sec) and $a=\Delta I/\Delta t$ is the slope at each point over Δt=1sec. For each curve, we derive SNR = 2.1, 1.9 and 2.2 for Δt = 1sec and obtain NEP = 1.9x10^{-18}, 3.4x10^{-18} and 3.3x10^{-18} W/Hz$^{1/2}$.

It is interesting if the noise in the present measurements arises from detector specific electrical noise or is ascribed to the photon noise. If the detector noise is ignored, random events of photon arrival give rise to the photon noise, like the shot noise in the electron current. On the average, $N\Delta t$ photons are counted over integration time of Δt when the photon count rate is $N = \eta \Phi = a/\Delta I_{Step}$. Since the standard deviation from the average is given by $(N\Delta t)^{1/2}$ in the Poisson distribution, SNR of an ideal detector (free from any detector-specific noises) is $(N\Delta t)^{1/2}$. It follows that NEP is given by (Knuse et al., 1962)

$$NEP_{BLIP} = h\nu \sqrt{\frac{2\Phi}{\eta}} = h\nu \frac{\sqrt{2N}}{\eta} \tag{6}$$

in the condition when the noise is dominated by the photon noise; i.e., in the background limited performance (BLIP). We have $NEP_{BLIP} = 1.9 \times 10^{-18}$, 2.5×10^{-18} and 2.6×10^{-18} W/Hz$^{1/2}$, respectively, for the three conditions ($N = 4.2$, 6.7 and 7.4) of the data in the inset of Fig.10. The fact that these values are close to the NEP values derived from Eq.(5) indicates that the fluctuation in the present measurements comes from the randomness in the photon arrival events and that the detector-specific noise is indiscernibly small.

The true NEP is obtained by replacing N in Eq.(6) by the dark count rate Γ, (Brule Technologies, Inc., 1980)

$$NEP = h\nu \frac{\sqrt{2\Gamma}}{\eta} \tag{7}$$

In the present spectrometer, weak background radiation makes hinders direct determination of Γ because of background blackbody corresponding at 50 K. In another optical system, we studied the detector in the complete dark condition, and obtained $\Gamma = 0.5$ sec^{-1}, which yields $NEP = 6.8 \times 10^{-19}$ W/Hz$^{1/2}$.

The fact hat the noise in the present measurements arises from the photon noise is also supported by the inter-step interval histograms. A time trace of ΔI for $P_{in} = 1$mW was studied for 700 seconds in the present spectrometer (partially shown in Fig. 10). The local slope of the curve, $a(t) = \partial I/\partial t$, was numerically calculated at each time point t of the curve, where the value is averaged over 1sec at each point. Figure 13 (a) represents the frequency of different slopes as a function of the steps per seconds or $a(t)/\Delta I_{Step}$ with $\Delta I_{Step} = 3$pA. In the trace of ΔI, upward stepwise change as well as downward change are recognized, which were separately analyzed in Fig.13 (b) (Gain) and (c) (Loss). As expected for single-photon detection, all histograms in Figs. 10 are fairly well described by the Poisson distribution, $f(k,N) = e^{-\nu} \nu^k/k!$, where k is the number of occurrence of an event and ν is the variance. We mention that the average rate of Loss events $\nu = 0.5$ in Fig.13 (c) is close to the dark count rate.

Specific detectivity, D^*, is the reciprocal value of NEP normalized by the detector active area S_{iso},

$$D^* = \frac{\sqrt{S_{iso}}}{NEP} \tag{8}$$

With $S_{iso} = 64 \times 10^{-8}$cm^2 for the present detector, we have $D^*_{BLIP} = 4.1 \times 10^{14}$, 2.3×10^{14} and 2.5×10^{14}W/Hz$^{1/2}$ in the background limited conditions ($NEP = 1.9 \times 10^{-18}$, 3.4×10^{-18} and 3.3×10^{-18} W/Hz$^{1/2}$). As for the true specific detectivity of the detector at 4.2K, we have

$D^* \cong 1.2 \times 10^{15} \text{cmHz}^{1/2}/\text{W}$ from $NEP \cong 6.8 \times 10^{-19}\text{W}/\text{Hz}^{1/2}$. CSIPs are thus highly sensitive detectors with values of D^* much higher than those of conventional detectors like QWIPs ($D^* = 10^{10} \text{-} 10^{13}\text{cmHz}^{1/2}/\text{W}$ at 4.2~80K) (Levine, 1993; Yao et al., 2000) and mercury-cadmium telluride (MCT) detectors ($D^* \sim 10^{10}\text{cmHz}^{1/2}/\text{W}$ at ~80K) (Rogalski, 2002).

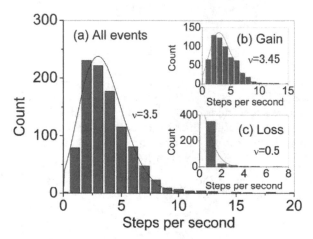

Fig. 13. Histogram of the photo-current ΔI per second. The photocurrent is measured in unit of ΔI_{Step} =3pA so that the horizontal axis is scaled by the number of steps per second, $(\Delta I / \Delta I_{Step}) / \Delta t = a / \Delta I_{Step}$. The data are derived by analyzing a time trace of ΔI for 700 sec at P_{in}=1mW: (a) All events of both gain signal (generation of holes) and loss signal (loss of holes due to recombination) are included. In (b) and (c), the gain signal and the loss signal are separately shown.

The intrinsic speed of detection will be limited by the transistor operation, which may be faster than 1 ns. In actual practice of detector operation, the detection speed may be determined by the frequency of applicable rest pulse, which may be restricted to about 300 MHz (~3 ns interval). The maximum incident radiation power will be hence $P_{max} \sim (7 \times 10^3 \, h\nu/3 \, ns)/\eta \sim 2 \, \mu\text{W}$, which makes us to expect that the dynamic range can be expanded to 10^{13} by assuming the minimum detectable power to be $P_{min} \sim 1 \times 10^{-19}\text{W}$.

7. Improvement of quantum efficiency

Quantum efficiency η is an important parameter for the detector. High values of η (>30%) are reported in conventional QW IR photodetectors, where photons are absorbed by more than 30 multiple QWs. High quantum efficiency is not trivial in CSIPs, where photons are absorbed by only one QW. In most experimental conditions of CSIP for λ = 12-15 μm, the efficiency $\eta = \eta_1 \eta_2$ is primarily determined by the coupling strength of electrons to the incident radiation, namely, $\eta \approx \eta_1$ ($\eta_2 \approx 1$).

Recently we proposed and demonstrated efficient photo-couplers for CSIPs (λ=15 μm) by exploiting surface-plasmon-polariton (SPP) resonance occurring in aperture metal sheets coated on top of the crystal surface (Fig. 14 (a)) (Nickels et al, 2010). The SPP resonance induces wavelength-selective strong electric confined near the surface of the metal sheets intensifying the subband transition in the QW 100 nm below the surface. The quantum efficiency has been experimentally studied by applying different metal photocouplers as displayed in Fig. 14 (b) and compared with simulation results in Fig. 14 (c).

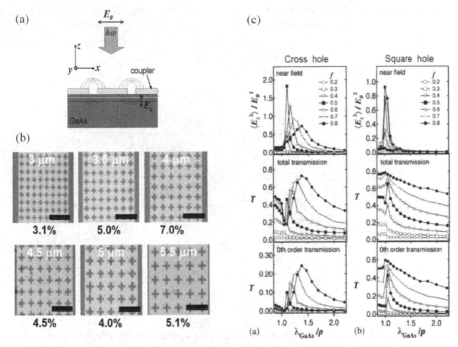

Fig. 14. (a) Schematic representation of photocoupler. (b) Metal meshes studied. White and black characters indicate, respectively, the lattice period p and the experimentally derived quantum efficiency η. (c)Results of simulation calculation are shown for (a) cross-hole arrays and (b) square-hole arrays: from the top, normalized field intensity $<E_z^2>/<E_{z0}^2>$, the total power transmission, and the zeroth-order transmission, against the ratio λ_{GaAs}/p with $\lambda_{GaAs} = 4.3\ \mu m$. E_z is the electric field normal to the QW plane at the position of upper QW(100 nm below the surface) and E_{x0} is the electric field of incident radiation. Theoretically, $<E_z^2>/<E_{z0}^2>$,= 2 gives roughly $\eta = 7\%$, which agrees with the experiment

In the experiments, an efficiency of $\eta \approx 7\%$ has been achieved by utilizing the surface plasmon excitation in a 2-D metal hole array (inductive metal mesh) with cross-shaped holes lining up at a lattice period of 4 μm (~ wavelength in GaAs). Other coupler geometries such as 2-D metal hole arrays with square-shaped holes, 2-D metal patch arrays (capacitive metal mesh), and patch antennas (microstrip antennas) have yielded lower values of $\eta \approx 1\%$–3% (Ueda et al., 2008; Nickels et al, 2010). The device used for Fig. 12, for instance, utilizes a patch antenna coupler yielding $\eta \approx 2\%$ (Ueda et al, 2008). Simulation suggests that $\eta \approx 7\%$ is not a highest achievable value but can be improved by a factor of ~3 ($\eta \approx 20\%$) when a resonant cavity is formed by the metal hole array and the n-type substrate. Values of the current responsivity and the NEP given in the section 6 have been so far derived by assuming $\eta = 2\%$. The figure-of-merits of the CSIP will be improved in accordance with η.

8. Temperature dependence of the performance

Higher temperature operation is desired for practical applications. There is in general, however, a trade-off between high sensitivity of an IR detector and operation temperature. The external constraint is that sensitive devices are saturated by strong background blackbody radiation from surrounding materials. The internal constraint is that the thermionic emission inside the device becomes equivalent to photo-emission. The former may be determined by optical setup, as well as relative intensity of signal radiation to the background. Here, we discuss the latter intrinsic constraint.

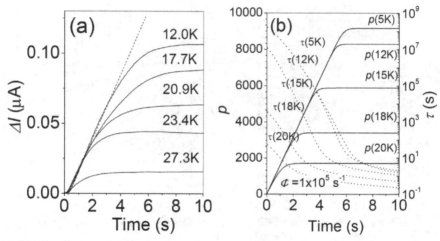

Fig. 15. (a) Time traces of the photocurrent with $\Phi=1\times10^5$ s^{-1} at different temperatures. (b) Theoretical time traces of photo current in terms of p at different T (solid line), and lifetime change (dotted line)

In Fig. 15 (a), time traces of photo-current at different temperatures are displayed under the fixed photon flux $\Phi=1\times10^5$ s^{-1}(Ueda et al., 2009). The temperature effect appears as the lower amplitude of photo-current saturation. It should be noted that the slope, $\alpha=\Delta I/\Delta t=\eta\Phi I_e$, in the initial stage of each trace is independent of T, assuring that ηI_e is independent of T. This means higher frequency reset operation is required, i.e., the integration time is shortened, in the elevated temperatures. The photo-signal is discernible up to 30K for the CSIP of $\lambda= 15$ µm. The derived NEP and D^* up to $T=23$ K with integration time of 1s are given as $NEP=8.3\times10^{-19}$ W/Hz$^{1/2}$, and $D^*=9.6\times10^{14}$ cm Hz$^{1/2}$ /W, which are not very different from the 4.2 K values mentioned above (Ueda et al., 2009).

The understanding of photo-current saturation directly leads to temperature dependent physics inside the device. The potential profile changing in the detection is shown in Fig. 5. The potential height of the triangular barrier in equilibrium is $U=\Delta E-\delta U$, i.e., barrier hight is lower, by $\delta U\approx15$ meV, than subband energy splitting of ΔE. Here the barrier hight $U=U_U=U_L$ is measured from electrochemical potential of upper or lower QW, $\zeta=\zeta_L=\zeta_U$.

Under illumination, the energy of the floating upper QW with p photoholes decreases against the grounded lower QW by $\Delta U= |\zeta_L-\zeta_U| = pe^2d / \kappa LW$, where e is the charge of electron, d is the distance between the upper and the lower QWs, and κ=12 x (8.85x10^{-12}) F/m is the electronic permittivity of the crystal. The net potential barrier height measured from ζ_L, then decreases to be U_L= (ΔE-δE)-ΔU by p photoholes, while U_U =ΔE-δE remains constant. The barrier reduction assists the electrons in the lower QW to recombine with photoholes in the upper QW. The photo-current saturation occurs when the recombination process becomes equivalent to the photohole generation process. Obviously, more electrons in the lower QW contribute to the recombination process at higher temperature. Finally at around T=30 K, the number of traveling electron across the barrier exceeds that of photo-emitted electrons even with no band-deformation.

Temperature dependence shown in Fig. 15 (a) arises from decrease of the lifetime of photoholes. The number of photoholes is determined by the rate equation (3). The recombination rate, p/τ, is given by difference between $\dot{n}(U_L)$ and $\dot{n}(U_U)$,

$$\frac{p}{\tau} = \dot{n}(U_L) - \dot{n}(U_U) \tag{9}$$

where $\dot{n}(U_U)$ is the rate from the upper QW to the lower QW and $\dot{n}(U_L)$ the rate of the opposite flow. Here again, U=U_L and U=U_U are the potential heights of triangular barrier measured from each electrochemical potential ζ of the lower QW, ζ=ζ_L, and the upper QW, ζ=ζ_U, respectively (Fig. 5).

The thermal electron emission is given as

$$\dot{n}(U) = \frac{LW}{d} D \int_0^\infty v\Theta f d\varepsilon \tag{10}$$

where D=$m^*/\pi\hbar^2$ is the two-dimensional density of states with m*=0.0665x(9.1x10^{-31})kg the effective mass of conduction electrons in GaAs, and v=$(2\varepsilon/m^*)^{1/2}$ the electron velocity with ε measured from the bottom of the conduction band, Θ is the transmission probability, and f= [1+exp{$(\varepsilon$-$\zeta)/k_BT$}]$^{-1}$ is the Fermi distribution function with k_B the the Boltzmann constant,. Here, we use WKB approximation of Θ, in which Θ=1 (at ε>U), and

$$\Theta = \exp\left[-\frac{4}{3eF}\left(\frac{2m^*}{\hbar^2}\right)^{\frac{1}{2}}(U-\varepsilon)^{\frac{3}{2}} \right] \quad (\text{at } \varepsilon<U), \tag{11}$$

where F= U_L/ed is the electric field given by the slope of the deformed triangular barrier.

As shown in Fig. 15 (b), the predicted time traces of p (photo-signal) at different T's substantially reproduce experimental results. It should be noted that adjustable parameters are not involved in the calculation. The temperature dependence is hence concluded to arise from those thermally activated electrons that are transmitted across the potential barrier to be recombined with holes in the isolated QW.

At the temperature limit, T_{lim}, Eqs. 3 and 9 becomes $2\eta\Phi$= $\dot{n}(U_L)$,by assuming $\dot{n}(U_U)\approx\eta\Phi$ under dp/dt=0 and ΔU=0. The relationship between subband energy the ΔE and temperature limit T_{lim} can be obtained:

$$\Delta E - \delta E \approx -k_B T_{\lim} \ln\left(\frac{2\eta \Phi d}{LWDvk_B T_{\lim}}\right) = \gamma T_{\lim} \tag{12}$$

where $v=(2E_F/m^*)^{1/2}$ is Fermi velocity. Here, integration is made only for $E > U$ with $\Theta=1$ and $f \approx \exp(-E/k_BT)$, i.e., only thermionic emission is considered. Since contribution of T_{lim} in the logarithm is not significant, the temperature limit shows almost linear dependence on the subband energy with quasi-constant $\gamma > 0$. For the $\lambda=15$ μm, the parameter values yields $T_{lim} =29$ K, which is close to the experimentally observed limit temperature.

9. Conclusion

We have developed the novel ultrasensitive detectors, named charge-sensisve infrared phototransistors (CSIPs). We demonstrated the single-photon detection in the wavelength range of 10 - 50 μm for the first time. The accurately determined specific detectivity as $D^*=9.6 \times 10^{14}$ cm Hz$^{1/2}$/W for $\lambda=14.7$ μm is by a few orders of magnitude higher than those of other conventional detectors. In addition, the planar FET structure of CSIPs is feasible for developing large scale array including monolithic integration with readout circuit. The quantum efficiency is improved up to 7% by using the surface-plasmon polariton couplers. The detectors can keep the sensitivity up to T=23 K for $\lambda=15$ μm, and the temperature dependence of the detector performance is reasonably interpreted by a simple theoretical model. We developed CSIPs for several wavelengths in 10-50μm, covering attractive spectral range for passive observation of spontaneous emission from RT object. Recently CSIPs have been applied to the construction of a highly sensitive passive microscope (Kajihara et al., 2009; 2010; 2011).

10. Acknowledgment

This work was supported by CREST and SORST projects of Japan Science and Technology Agency (JST). The reserach work has been carrierd with number of collaborators. Particularly T. Ueda would like to thank Z. An, P. Nickels, N. Nagai, Z. Wang and S. Matsuda for their valuable contributions.

11. References

An, Z., Ueda, T., Komiyama, S. & Hirakawa, K. (2007). Reset Operation of Qwantum-Well Infrared Phototransistors, *IEEE Transaction on Electron devices*, Vol. 75, pp. 1776-1780, ISSN 0018-9383

Andersson, J. & Lundqvist, L. (1991). Near-Unity Quantum Efficiency of AlGaAs/GaAs Quantum Well Infrared Detectors Using a Waveguide with a Doubly Periodic Grating Coupler, *Applied Physics Letters*, Vol. 59, pp. 857 1-3, ISSN 0003-6951

Astaviev. O., Komiyama, S., Kutsuwa T., Antonov, V. & Kawaguchi, Y. (2002). Single-Photon Detector in the Microwave range, *Applied Physics Letters*, Vol. 80, pp. 4250 1-3, ISSN 0003-6951

Burle Technologies, Inc. (1980). *Photomultiplier Handbook – Theory Design Application*, Lancaster, PA

Cwik, T. & Yeh, C. (1999). Higjly Sensitive Quantum Well Infrared Photodetectors, *Journal of Applied Physics*, Vol. 86, pp. 2779 1-6, ISSN 0021-8979

Day, P., LeDuc, L., Mazin, B., Vayonakis, A. & Zmuidzinas, J. (2003). A Broadband Superconducting Detector Suitable for Use in Large Arrays, *Nature*, Vol. 425, pp. 817-821, ISSN 0028-0836

Hashiba, H., Antonov, V., Kulik, L., Tzalenchuk, A., Kleindschmid, P., Giblin, S. & Komiyama, S. (2006). Isolated quantum dot in application to terahertz photon counting, *Physical Review B*, Vol. 73, pp. 081310 1-4, ISSN 0613-1829

Ikushma, K., Yoshimura, Y., Hasegawa, H., Komiyama, S., Ueda, T. & Hirakawa, K. (2006). Photon-Counting Microscopy of Terahertz Radiation, *Applied Physics Letters*, Vol. 88, pp. 152110 1-3, ISSN 0003-6951

Komiyama, S., Astaviev. O., Antonov, V., Kutsuwa T. & Hirai, H. (2000). A Single-Photon Detector in the Far-Infrared Range, *Nature*, Vol. 403, pp. 405-407, ISSN 0028-0836

Kajihara, Y., Komiyama, S., Nickels, P. & Ueda, T. (2009). A Passive Long-Wavelength infrared Microscope with a Highly sensitive phototransistor, *Review of Scientific Instruments*, Vol. 80, pp. 063702 1-4, ISSN 0034-6748

Kajihara, Y., Kosaka, K., Komiyama, S., Kutsuwa T. & Hirai, H. (2010). A Sensitive Near-Field Microscope for Thermal Radiation, *Review of Scientific Instruments*, Vol. 81, pp. 033706 1-4, ISSN 0034-6748

Kajihara, Y., Kosaka, K.& Komiyama, S. (2011). Thermally excited Near-Field Radiation and Far-Field Interfarence, *Optics Express*, Vol. 19, pp. 7695-7704, ISSN 1094-4087

Knuse, P., McGlauchlin, L. & McQuistan, R. (1962). *Elements of Infrared Technology*, Wiley, ISBN 978-047-1508-86-1, New York

Levine, B. (1993). Qwantum-Well Infrared Photodetectors, *Journal of Applied Physics*, Vol. 74, pp. R1-81, ISSN 0021-8979

Nakagawa, T. & Murakami, H. (2007). Mid- and far-infrared astronomy mission SPICA, *Advances in Space Research*, Vol. 40, pp. 679-683, ISSN 0273-1177

Prochazka, I. (2005). Semiconducting Single Photon Detectors: The State of the Art, *Physica Status Solidi (c)*, Vol. 2, pp. 1524-1532, ISSN 1610-1634

Rogalski, A. & Chrzanowski, K. (2002). Infrared Devices and Techniques, *Opt-Electrics Review*, Vol. 10, pp. 111-136, ISSN 1230-3402

Schoelkopf, R., Moseley, S., Stahle, C., Wahlgren, P. & Delsing, P. (1999). A concept for a submillimeter-wave single-photon counter, *IEEE Transactions on Applied Superconductivity*, Vol. 9, pp. 2935-2939, ISSN 1051-8223

Tidrow, M. (2000). Device Physics and State-of-the-art of Quantum Well Infrared Photodetectors and Arrays, *Materials Science and Engineering: B*, Vol. 74, pp. 45-51, ISSN 0921-5107

Ueda, T., An, Z., Hirakawa, K. & Komiyama, S. (2008). Charge-Sensitive Infrared Phototransistors: Characterization by an All-Cryogenic Spectrometer, *Journal of Applied Physics*, Vol. 103, pp. 093109 1-7, ISSN 0021-8979

Ueda, T., An, Z., Hirakawa, K., Nagai, N. & Komiyama, S. (2009). Temperature Dependence of the Performance of Charge-Sensitive Infrared Phototransistors, *Journal of Applied Physics*, Vol. 105, pp. 064517 1-8, ISSN 0021-8979

Ueda, T., Soh, Y., Nagai, N., Komiyama, S. & Kubota, H (2011). Charge-Sensitive Infrared Phototransistors Developped in the Wavelength Range of 10-50μm *Japanese Journal of Applied Physics*, Vol. 50, pp. 020208 1-3, ISSN 0021-4922

Wei, J., Olaya, D., Karasik, B., Pereverzev, S., Sergeev, V. & Gershenson, M. (2008). Ultrasensitive hot-electron nanobolometer for Terahertz Astrophysics, *Nature Nanotechnology*, Vol. 3, pp. 495-500, ISSN 1748-3387

Yao, J., Tsui, D. & Choi, K. (2000). Noise characteristics of quantum-well infrared photodetectors at low temperatures, *Applied Physics Letters*, Vol. 76, pp. 206 1-3, ISSN 0003-6951

Nickels, P., Matsuda, S., Ueda, T., An, Z. & Komiyama, S. (2010). Metal Hole Arrays as a Resonant Photo-Coupler for Charge-Sensitive Infrared Phototransistors, *IEEE journal of Quantum Electronics*, Vol.46, pp. 384-3906, ISSN 0018-9197

8

Geiger-Mode Avalanche Photodiodes in Standard CMOS Technologies

Anna Vilà, Anna Arbat, Eva Vilella and Angel Dieguez
Electronics Department, University of Barcelona
Spain

1. Introduction

Photodiodes are the simplest but most versatile semiconductor optoelectronic devices. They can be used for direct detection of light, of soft X and gamma rays, and of particles such as electrons or neutrons. For many years, the sensors of choice for most research and industrial applications needing photon counting or timing have been vacuum-based devices such as Photo-Multiplier Tubes, PMT, and Micro-Channel Plates, MCP (Renker, 2004). Although these photodetectors provide good sensitivity, noise and timing characteristics, they still suffer from limitations owing to their large power consumption, high operation voltages and sensitivity to magnetic fields, as well as they are still bulky, fragile and expensive. New approaches to high-sensitivity imagers tend to use CCD cameras coupled with either MCP Image Intensifiers, I-CCDs, or Electron Multipliers, EM-CCDs (Dussault & Hoess, 2004), but they still have limited performances in extreme time-resolved measurements.

A fully solid-state solution can improve design flexibility, cost, miniaturization, integration density, reliability and signal processing capabilities in photodetectors. In particular, Single-Photon Avalanche Diodes, SPADs, fabricated by conventional planar technology on silicon can be used as particle (Stapels et al., 2007) and photon (Ghioni et al., 2007) detectors with high intrinsic gain and speed. These SPAD are silicon Avalanche PhotoDiodes biased above breakdown. This operation regime, known as Geiger mode, gives excellent single-photon sensitivity thanks to the avalanche caused by impact ionization of the photogenerated carriers (Cova et al., 1996). The number of carriers generated as a result of the absorption of a single photon determines the optical gain of the device, which in the case of SPADs may be virtually infinite.

The basic concepts concerning the behaviour of G-APDs and the physical processes taking place during their operation will be reviewed next, as well as the main performance parameters and noise sources.

1.1 Basic concepts for Geiger-mode avalanche photodiodes, G-APDs

APDs can be obtained by implementing two possible approximations that produce two different structures with differentiated capabilities. On one hand, thin silicon APDs (Lacaita et al., 1989) are devices with a depletion layer of few micrometres and low breakdown voltages. They also present good detection efficiency and time resolution. As in planar

APDs it is important to avoid the possibility of edge breakdown of the sensor, many different terminations have been proposed, generating a variety of planar CMOS compatible devices. On the other hand, thick APDs (Cova et al., 2004) are devices with a depletion layer of some tens of micrometres that work at high breakdown voltage; they have good detection efficiency but moderate time resolution. These components are fabricated in dedicated technologies, increasing their cost.

An example of these two different structures is shown in Fig. 1.A: a thick APD at the left side (Alexander, 1997) and a thin APD at the right side. The structures are composed of different sections of semiconductor with different doping profiles shown in B. In C, the charge density of the structures is represented. The p region develops a net negative charge, while the n region develops a net positive charge, generating the energy diagrams visible in D and an electric field profile such as in E.

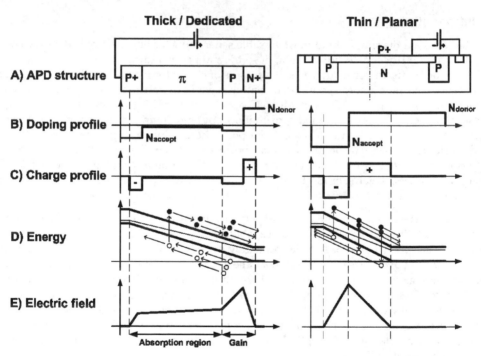

Fig. 1. Comparison between thick (left) and thin (right) APD structures. A) structure, B) doping profile across the dash-dotted line in A, C) charge profile, D) energy-band diagram, E) electric field profile.

An electron-hole pair, EHP, can be generated due to photon absorption, minimum-ionizing particle crossing the detector or by thermal generation. Due to the electric field in the sensing area, carriers are accelerated to the collection zone. The high electric field give to the carriers enough kinetic energy to produce additional EHP by impact ionization. This additional charge is also accelerated and eventually produces new EHP in their way towards the electrodes. The whole effect generates a multiplication of the electric signal from a single photon.

There are two possible operation modes depending on the biasing of the APD: linear (Fig. 2.A) and Geiger (Fig. 2.B). In the linear mode, only the electrons achieve enough kinetic energy to produce new EHPs by impact ionization, producing gains around 1000 (Kindt, 1994). The gain is then moderate and affected by strong excess noise (McIntyre, 1972), making difficult the detection of single photons. The gain is strongly dependent on the biasing voltage, requiring a uniform field over the sensible area to have a constant gain. When the device is biased well above the breakdown voltage, Geiger mode, the electric field in the depletion layer is so high that also holes produce new EHPs, generating a self-sustaining avalanche current. In this case, the avalanche can be triggered by a single photon or a single EHP generated by a MIP, and the device becomes a SPAD. The current rises swiftly (sub-nanosecond rise-time) to a macroscopic steady level in the milliampere range, irrespective to the quantity of photons detected (essential effect for silicon photomultipliers, where the output signal is proportional to the number of APDs in avalanche).

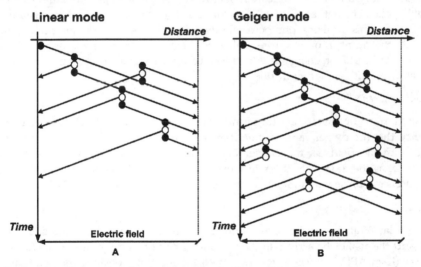

Fig. 2. Avalanche phenomena due to impact ionization. A) In linear mode, only electrons contribute to EHP generation. B) In Geiger mode, both electrons and holes produce newEHPs, promoting a self-sustained avalanche.

In Geiger mode, the current generated during the avalanche continues to flow until lowering the bias voltage down to the breakdown quenches it. Then, the lower electric field is no longer able to accelerate carriers to impact-ionize lattice atoms, therefore current ceases. In order to be able to detect another photon, the biasing needs to be restored to the initial value in a recharge process to leave the detector ready to work again.

Consequently, the detection cycle includes the quenching and the recharge periods, and usually specific circuitry helps to do these actions. The basic goal is to limit the avalanche charge flowing through the APD at every ignition, in order to reduce three detrimental effects:

- Afterpulsing: avalanche events produced by the release of charges trapped into deep levels (Cova et al., 1996).

-   Self-heating: more dark counts are generated, but also breakdown voltage is increased and excess voltage decreased, reducing the detection efficiency.
-   Crosstalk: interference between neighbouring sensors produced by emission of secondary photons (Zappa et al., 1997).

1.1.1 Performance parameters

In this context, the main figures of merit that have been used to describe the capabilities of G-APDs as photodetectors are shortly reviewed next. Some of them are specific of the case under study and have no sense related to other types of technologies. However, all they are of crucial importance in G-APDs.

Gain

First of all, the gain or multiplication factor, M, is defined as the ratio between the observable photocurrent at the APD terminals and the internal photocurrent before multiplication. This primary photocurrent depends on the detector quantum efficiency, defined as the number of electrons collected in a pixel over the number of photons penetrating it, the photon energy and the received optical signal power. This effect is similar to that achieved with a photomultiplier.

Ionization coefficients

The ionization coefficients for electrons and holes, a_e and a_h respectively, are defined as the probability that the carrier causes some ionization within a unit length. These coefficients increase with the field, as the carrier achieves a higher energy, while decreases with temperature, as the atoms are thermally excited, reducing the collision probability. High-quality APDs can present a ratio between a_h and a_e of between 0.003 and 0.01.

Time delay

The time delay of an avalanche corresponds to the time between the photodiode generates an EHP and the avalanche arrives to the measuring circuit, and it represents a fundamental parameter of an APD because it, together with the RC time-constant and the transit time, determines its bandwith. The time delay is determined by the transition times for the electrons/holes to reach the avalanche region (i.e. the time required for charges to go through an absorption region and arrive to the avalanche region) and the avalanche build-up time.

Timing resolution or jitter

The timing resolution or timing jitter of the sensor is defined as the statistical fluctuation of the time interval between the arrival of the photons at the sensor and the output-pulse leading edge. It can be described as the full-width half-maximum measure of the temporal variation in the avalanche breakdown pulses resulting from an incident photon beam. Among the timing-resolution components are the variation caused by the generated-carrier transit time from the depletion layer to the multiplication region, which is dependent on the depth of absorption of the incident photon and, more important, the statistical build up of the avalanche current itself (Ghioni et al., 1988), which timing improves with the excess bias voltage.

Photon detection efficiency, PDE, or photon detection probability, PDP

The photon detection efficiency, PDE, is the probability of avalanche generation for each detected photon. For a photon to be detected, it must be absorbed in the detector active volume and generate a primary carrier, but it is also necessary for this primary carrier to succeed in triggering an avalanche. The photon detection probability, PDP, is related to the PDE, and it is defined as the probability of a photon to generate an avalanche. Both concepts are wavelength dependent and increase with the excess bias voltage, as the higher electric field enhances the triggering probability.

Sensitivity or dynamic range

The dynamic range, DR, quantifies the ability of a sensor to adequately image both clear lights and dark shadows in the same scene. It is defined as the maximum signal divided by the noise. In the case of APDs, the DR is given by the maximum number of counts generated by the sensor and the dark count noise. The sensitivity increases with the biasing voltage and depends on the technology.

Fill factor

The fill factor is the ratio between the light-sensitive area of a pixel and its total area. The electric field strength should be consistent across the whole active area (the central photon-sensitive portion of the detector), so as to yield a homogeneous breakdown probability. However, electric field concentration at the edges and corners can cause higher photon detection probability and dark counts. Such zones should be avoided by proper design solutions such as guard rings or shallow trenches, which limit the active detector volume. This is a main drawback for present G-APDs, especially for large-area high-precision applications such as particle detectors in colliders or telescopes.

Dead time

The dead time is the time during which the sensor is not responsive to further incoming photons, due primarily to accumulation of charge in the active region. It involves the periods comprising the avalanche quenching and the device recharge, i.e. the reset of the final bias conditions. However, in the case of a passively quenched SPAD this is not strictly the case, because, as the device is recharged via the quenching resistor, it becomes increasingly biased beyond its breakdown voltage, so it is able to detect the next photon arrival prior to being fully reset. This behaviour is coupled with a significant fluctuation in the reset waveform and can increase the afterpulsing probability due to inadequate trap flushing time.

1.1.2 Noise sources

The inherent noise sources of APDs are dark counts and afterpulsing events. There is also an external noise source between neighbouring sensors due to electrical and optical crosstalk.

Dark count rate, DCR

A dark count is an avalanche event caused by non-photogenerated carriers, which can be originated from four factors (Haitz, 1965): diffusion from neutral regions, thermal generation, band-to-band tunnelling or by release from a charge trap (see also afterpulsing).

This generation is critical when operating in Geiger mode, as it can produce an output pulse indistinguishable from the one originated by a detection event. The per-second rate at which dark counts occur is the DCR in Hertz.

Dark-count noise increases with temperature, as the thermal carrier generation increases and so the probability of avalanche. It also increases with the excess voltage because of two effects, namely the field-assisted enhancement of the emission rate from generation centres and the increase of the avalanche-triggering probability with the higher electric field at the depletion layer, which reduces the recombination probability. Of course, thermal generation is strongly dependent on the fabrication process, which determines the concentration of traps, the breakdown voltage and the avalanche probability. On the contrary, tunnelling generation depends on the doping profile and the bias voltage, and its temperature dependence is weak.

Afterpulsing

Charge traps due to unintentional impurities and crystal defects can result in generation-recombination centres. The high current peak through the junction during an avalanche breakdown introduces a probability that the trap is filled by a carrier that is later released, initiating a second, follow-on Geiger 'after-pulse' (Haitz, 1965). Afterpulsing is then a false detection caused by secondary dark pulses correlated to a previous primary one. This phenomenon suffers from a statistical fluctuating delay whose mean value depends on the deep levels actually involved.

The number of carriers captured during an avalanche pulse increases with the total number of carriers crossing the junction, that is, with the total charge involved in the avalanche pulse, and with the number of traps in the multiplication area. Therefore, afterpulsing increases with the delay of avalanche quenching and with the current intensity, which is proportional to the excess bias voltage. This overvoltage also enlarges the depletion zone, and so the number of available traps, increasing the trapping probability. Finally, released charges can retrigger avalanche.

Crosstalk

A crosstalk is a signal interference between neighbouring pixels produced by electrical or optical interaction between them. The electrical crosstalk corresponds to the migration of carriers from one pixel to another, and can be probably reduced by a potential barrier designed properly to collect these charges. On the other hand, hot carriers in an avalanche emit 2.9 photons per 10^5 electrons (Lacaita et al., 1993) that can produce an avalanche in the neighbouring pixel generating an optical crosstalk. The probability of this phenomenon depends on many factors, such as pixel size, dead space between active areas, sensitive volume and gain, but it can be diminished by reducing the number of carriers generated in the avalanche.

2. Technology considerations for GAPDs

This section consists on a review of the fundamental technology considerations that deserve to be taken into account referred to the actual G-APDs, including the state-of-the-art of avalanche photodiodes fabricated in CMOS compatible technologies.

2.1 Historical review of APDs technologies

The switching effect observed in reverse-biased silicon p-n junctions by McKay (McKay, 1954), marked the beginning on the study of the structures in avalanche. Imperfect junctions with a large number of impurities were used in initial investigations. In 1961 Shockley (Shockley, 1961) proposed the ideal reverse junction characteristics which were confirmed in the early 1960s (Goetzberg et al., 1963), (Haitz, 1964) and (Haitz, 1965). With initial progresses of silicon technology, it was possible to fabricate better junctions nearly free from defects (called microplasmas by then). But practical devices were large (about $1mm^2$) and they did not attract much attention due to their high cost low efficiency.

In the early 70s, McIntyre (McIntyre, 1972) and Webb (Webb et al., 1974) developed new APD structures that were the precursors of commercial devices implemented in the Single Photon Counting Modules, SPCM. These new devices were fabricated on a sophisticate dedicated process which was incompatible with the planar technology (thick-APDs), but improved the photodetection efficiency. This was the beginning of dedicated technologies for APDs, able to reduce the dark count noise and improve the features of the detector.

Evolution has allowed APDs exhibiting excellent quantum efficiency, with values around 80% in the near ultra-violet range, dropping to about 40% in the blue region, which is to be compared to typical values of 5-8 % in the blue for standard photomultipliers. Although in the late 80s the planar technology to fabricate APDs was improved by Ghioni when they introduced the first epitaxial device structure (Ghioni et al., 1988), nowadays mainly all commercial APDs are produced by dedicated technologies not compatible with CMOS planar technology, what increases the difficulty to integrate an array of detectors together with the control electronics.

At the beginning of the XXI century the research on APDs was reoriented to the integration of the sensors together with the electronics thanks to the improvement on the CMOS technologies. Main objectives were the fabrication of SPAD integrated cameras (Charbon, 2007) and of large SPADs, which were restricted to the dedicated technologies (Ghioni et al., 2007). The first APDs in CMOS compatible technologies were published by research centers (Spinelli et al., 1998, Jackson et al., 2001 and Gulinatti et al., 2005). They were developed using the high voltage CMOS, HV-CMOS, technologies (Rochas et al., 2002 and Zappa et al., 2004), as they provide a relatively low-doped deep n-well that allows up to 50 V of isolation from the substrate (Rochas et al., 2003a) and low noise detectors.

The first SPAD-based pixel array in CMOS technology has been demonstrated a few years ago (Rochas et al., 2003b), consisting on a 4 by 8 array of SPADs with a passive quenching circuit and a simple comparator (inverter). Further developments followed quickly, facilitated by the availability of commercial HV-CMOS technologies and of specially tailored imaging processes boosted by the huge market of mobile phone cameras. Designs using a 0.8 µm CMOS technology have also been presented. In (Niclass & Charbon, 2005) a 64x64 image sensor with single pixel readout is employed for 3D imaging. Another APD array based on event-driven reading was also developed using the same technology (Niclass et al., 2006a). The evolution of the technologies also produces an evolution on the sensors, and some arrays were presented using 0.35 µm technology (Sergio et al., 2007). An array of 128x2 APDs was developed including latches to preserve the timing information. Finally the evolution of APDs arrived to 130 nm, with an array of 32x32 sensors (Guerrieri et al., 2009).

In the last years, the use of submicronic technologies has also included the study of APD structures, and some comparison between technologies has been reported recently (Arbat et al., 2010a), concluding that high-voltage technology has a low dark-count rate related to the low trap concentration but the sensor is slow, while high-integration technology generates fast sensors with high dark count rates. In 2006 Finkelstein presented a new structure for an APD based on shallow-trench isolation, STI (Finkelstein et al., 2006a). Although a reduction of size, the structure suffered from many dark counts that limited the device performances. In 2007, an n-well-based structure was designed on a 130nm technology (Niclass et al., 2007a and b), and the modifications needed for the 130nm-process to improve the detector features were presented (Gersbach et al., 2008). With the same 130nm technology, p-guard STI structures were also proposed (Gersbach et al., 2009), as well as a lower p-doping well in a deep n-well (Richardson et al., 2009). Going further, the first APD structures obtained using 90nm technology have been reported (Karami et al., 2010). In a parallel way, the use of nanostructures of germanium deposited in silicon to obtain small APDs able to work at voltages of only 1.5V and speeds of 30GHz have been proposed (Assefa et al., 2010). Even, a novel effort to use APDs to construct detectors for charged particles has recently been reported (Graugés et al., 2010, Vilà et al., 2011). Results suggest that, through control of the doping concentration, devices with a much improved fill factor could be achieved. Consequently, submicron CMOS technologies with ever decreasing minimum feature size seem to be useful for SPAD fabrication.

2.2 Dedicated vs. standard technologies

The present situation of SPADs may be described in terms of the manufacturing processes and construction details. In particular, two main process categories can be distinguished:

a. Dedicated technologies, which improve performances by optimizing individual technological parameters. They generate low-noise high-quantum-efficiency sensors via thick depletion layers (> 30μm), but integration of electronics with sensor is impossible.
b. CMOS compatible technologies, which can be only compatible (i.e. full-custom processes optimized to yield the best possible performing single detector element) or standard CMOS processes, without any modifications to the layers normally available to the designers. Both single detectors and arrays have been implemented with HV-CMOS processes (Rochas et al., 2003a and b, Stoppa et al., 2007 and Vilà et al., 2011), demonstrating their suitability for SPADs.

Both types of APD processes offer different features. In Table 1, a comparative of the main characteristics of each type of sensors is given. As APDs are mainly used as photon detectors, the detection efficiency is given for wavelengths. When high-energy particles are considered, the detection efficiency is related to the number of electrons generated by ionization in the multiplication region due to the crossing particle.

A method to build a thick APD using dedicated technologies consists on building the sensor on a p substrate with an n+ layer on one side and a p+ diffused layer to improve the ohmic contact on the other side (Webb et al. 1974). This produces a reach-through structure formed by four layers in a vertical structure, generating a p+-π-p-n+ stack (see figure 1) that can be improved through processes covered by patents (McIntyre, 1990 and 1996). As its operation is based on the complete depletion of the device (30 to 100μm deep), the required voltages are high, typically over 100V. This produces sensors with low noise and high quantum efficiency (Cova et al., 2004, Lacaita et al., 1988 and Ghioni & Ripamonti, 1991). This technology allows the possibility to produce wide active areas, up to 500 µm in diameter, with variable depletion-layer thickness (Dautet et al., 1993).

Main features	Planar – Thin	Dedicated - Thick
Breakdown	10 – 50 V	100 – 500 V
Multiplication thickness	1 – 2 µm	30 – 100 µm
Active area diameter	5 – 150 µm	100 – 500 µm
Photon Detection Efficiency (PDE)	~ 45 % @ 500 nm ~ 32 % @ 630 nm ~ 15 % @ 730 nm ~ 10 % @ 830 nm 0.1 % @ 1064 nm	> 50 % @ 540 - 850 nm ~ 3 % @ 1064 nm
Resolution in photon timing	<100 ps FWHM	< 350 ps reach-through ~ 150 ps smoother field profile
Dark count rate	Technology dependent	Very low noise
Power dissipation	Low, cooling system not required	High, cooling system required
Array fabrication	Compatible	Not compatible
Robustness	Good	Delicated
Cost	Low	High

Table 1. Dedicated vs. planar technologies comparison.

There are some commercial compact modules that include the bias and the quenching circuitry (Spinelli et al., 1996 and EG&G, 1996). Some performances of commercial APDs based in reach-through structures are given in Table 2. Although the reach-through structures have good performance, they also have a number of practical drawbacks. The main problem is the high power dissipation due to the high biasing voltage, arriving to values of 5 to 10 W. Another negative aspect is the high fabrication cost of the sensors, due to a low fabrication yield. Finally, the produced devices are delicate and degradable, and of course not integrable with circuitry.

Feature	InGaAs	Si	Array 4x8
Active area (mm)	0.04	0.2 - 5	1.6 x 1.6 Pitch 2.3mm
Wavelength (nm)	950 - 1700	200 - 1100	320 - 1000
λ peak (nm)	1550	620 – 940	600
Sensibility @ λp (A/W)	0.8 – 0.9	0.42 -0.5	-
Quantum efficiency (%)	-	75 - 80	70
V breakdown (V)	40 - 60	150 - 300	400 - 500

Table 2. Hammamatsu commercial APD features.

APD devices developed in planar technologies generate a thin depletion layer of few micrometers, being called thin APDs. These devices are generated by epitaxy over silicon wafers (Spinelli et al., 1998). Over the last years, improvements of the technology to fabricate integrated circuits have allowed the development of APDs integrated in commercial CMOS

Reference	Technolgy (termination)	Diameter (µm)	V_{BD} (V)	DCR (c/s)	PDE (%) @500nm	FWHM (ps)
(Spinelli et al., 1998)	Double junction	10	27	-	35	35
(Jackson et al., 2001) (Jackson et al., 2002)	1.5µmNMRC (1)	15/20/30 40/50	30	200/1k/3k 9k/30k	26	-
(Gulinatti et al., 2005) (Ghioni et al., 2006)	CNR-IMM	20/50 100/200	24	300/700 4k/40k	38	30/30 31/35
(Rochas et al., 2002)	HV0.8µm (4)	80	19.5	-	-	-
(Rochas et al., 2003a) (Stapels et al., 2006)	HV 0.8µm (2)	7	25	900	30	60
(Zappa et al., 2004) (Zappa et al., 2005)	HV 0.8µm (2)	12	16	600	38	36
(Panchieri & Stoppa, 2007)	HV 0.7µm (2)	10x10 20x20 40x40	20.5 - 21	<10k <60k <200k	26	144
(Dandin et al., 2007)	0.5µm (5)	-	16.85	16k	-	-
(Mosconi et al., 2006)	HV 0.35µm (2 + IC)	20x20	28.3	3 – 5k	10	80
(Xiao et al., 2007)	HV 0.35µm (3)	-	45 - 50	<50	33	80
(Tisa et al., 2008)	HV 0.35µm (2 + IC)	20	24	<5k	37	39
(Niclass et al., 2006b) (Niclass et al., 2009)	0.35µm (2)	4/10	-	6/750	36	80
(Arbat et al., 2010a)	HV 0.35µm (2)	20x20	17.3	3.4k	-	-
(Finkelstein et al., 2006a) (Finkelstein et al., 2006b)	180nm IBM (6)	2x2 14x14	10	185k	12 - 20	-
(Faramarzpour et al., 2008)	180nm (2)	10/20	10.2	60k(10) 240k (20)	45	-
(Niclass et al., 2007a) (Niclass et al., 2007b)	130nm mod (2)	10	10	100k	41	144
(Gersbach et al., 2008) (Gersbach et al., 2009)	130nm mod (6)	4.3	10/17.2	1M/90k	36	125
(Richardson et al., 2009)	130nm STM (3)	8	14.4	<20	28	200
(Arbat et al., 2010b)	130nm STM (2)	20x20	10.5	3.5k	-	-
(Arbat et al., 2010b)	130nm STM (6)	20x20	10.55	1200k	-	-
(Karami et al., 2010)	90nm mod (2)	8x8	10	8.1k (0.13V)	8 - 14	398

Table 3. CMOS planar APDs. Termination (see section 2.3): 1-N+diffusion over p-well. 2-P-guard ring. 3-Deep p-well. 4-P-diffusion. 5-P-guard and polysilicon control gate. 6-Shallow trench isolation

technologies. Table 3 contains a summary of the published structures fabricated using planar technologies. See section 2-3 for details about termination. Some performances are associated to these devices, such as the possibility of integrating the electronics of the detector, of developing arrays of sensors, and of using low voltages and reducing the power consumption of the devices. One of the main disadvantages of the utilization of standard fabrication processes is the impossibility of modifying the fabrication process without the foundry interaction, inhibiting the possibility of device optimization.

2.3 Termination techniques for planar APDs

As commented before, the working principle of APDs requires a uniform electric field in the whole active area, what is only achieved in the central part of the sensor. Electrical-field concentration at the edges of the active area disturbs this uniformity and can cause premature edge breakdown, PEB. Different techniques have been developed in order to avoid PEB and consequently the device failure (Charbon, 2008), with the objective of limiting the electric field at the edges to be weaker than at the central multiplication region. Some of the most used options are summarized next.

Adding an n+ layer between the n substrate and the p surface confines the high electric field at the central part of the sensing area (Fig. 3), owing to the higher doping at the central junction (p-n+ instead of p-n at the edges). The concept was used with a p enrichment below the n+ diffusion to confine the electric field (Spinelli et al., 1998 and Jackson et al., 2000).

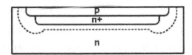

Fig. 3. N+ layer added between the n substrate and the p surface to avoid PEB. Termination technique (1) in table 3.

A second option largely used consists on the implementation of a p guard ring with lower doping level around the structure (Niclass et al., 2006b and Stoppa et al., 2005). This structure allows the possibility to share n-tubs between different sensors, reducing the pitch (Pancheri & Stoppa, 2007), and can be used in triple-well technologies which provide low-doping layers, as in the case of HV-CMOS (Xiao et al., 2007). This last approximation avoids heavily doped layers which are prone to have generation and trapping centres. Fig. 4 shows the initial and triple-well structures. Sensor arrays have been demonstrated with this technology (Tisa et al., 2008 and Arbat et al., 2010a) and it suitability has also been demonstrated in submicron technologies (Faramarzpour et al., 2008), but it is always important to avoid the full depletion of the n well, which would lead to punch through between the well and the substrate.

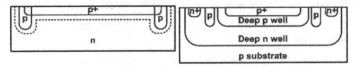

Fig. 4. Structures for avoiding PEB based on p-well guard ring at left, termination technique (2) in table 3, and deep tubs available in HV-CMOS technologies at right, termination (3) in table 3.

A third option frequently used consists on a p diffusion close to the edge of the p well at the sensing area (Rochas et al., 2002). In this case, the ion lateral diffusion during the processing generates a low-doped n region between p zones which can be completely depleted thanks to the biasing of an additional gate located on it (fig. 5, left). When using a p-substrate technology with doble well, it is possible to build similar guard-ring structures (fig. 5, right) with adjacent n wells placed at the minimum distance allowed by the technology (Dandin et al., 2007).

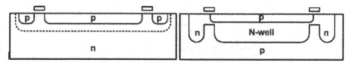

Fig. 5. Double low-doped n-well structure with gates (left) and guard rings and control gate signal (right). Termination techniques (4) and (5) in table 3, respectively.

Finally, for technologies beyond 250nm, the use of shallow trench isolation, STI, becomes a technological solution to reduce the negative effects of the birds-beak-shape of the SiO_2 (Finkelstein et al., 2006a). The STI prevents punch-through and latch-up, and it confines the electric field because the dielectric strength of SiO_2 is much higher than that of Si (fig. 6).

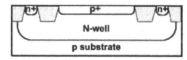

Fig. 6. Structure to avoid PEB based on STI elements. Termination technique (6) in table 3.

All these strategies have demonstrated suitability for preventing PEB, but not all of them are possible in a specific technology. The better choice depends on the possible structures.

3. G-APD simulation and characterization

Characterization of the individual sensing elements is needed for detector design, using different test benches and a black Faraday box. This section analyzes firstly the possibility of physical description of the device, to better understand its optical and electrical characteristics. Next, the characterization techniques usual for APDs are introduced: the I-V characteristic of the detector gives information about the breakdown voltage, the dark current below breakdown (linear mode) and the avalanche current. A 4-wire setup directly connected to the sensor terminals is recommended. Finally, the noise figures need to be characterized: dark count rate, afterpulsing and optical and electrical crosstalk. These parameters are not evident to minimize, and a definitive reading of actual detection requires new strategies based mainly on the accompanying electronics, as exposed in next section.

3.1 Physical simulation

Three main aspects have to be taken into account to calculate the electrical behaviour of the G-APDs: firstly, the charges generated in the semiconductor during the breakdown phenomenon; secondly, the movement of these charges inside the device towards its electrodes; and thirdly, the influence of the external circuitry on the recorded signal after detection, which is the only result that can be compared with experimental data.

Interaction between radiation and matter is a statistical phenomenon that can be calculated via Monte Carlo models able to describe a particle or photon beam as it interacts with the different layers in a device. The conventional model that explains the generation-recombination processes via traps is the well known Shockley-Read-Hall model (Shockley & Read, 1952), in which four basic mechanisms are supposed to be involved: electrons and holes capture and electrons and holes escape from a trap. However, for high electric field, tunnelling effects are dominant, and newer models are introducing corrections to take them into account.

So long, Monte Carlo calculations have then allowed a detailed description of a beam inside a device, discriminating the contribution of tunnelling to the dark-count rate, and explaining most of the inner effects of absorption, backscattering, etc. referred to a given structure (Vilà et al., 2011). After that, the evolution of the generated charges must be described, and classical drift-diffusion models can give some macroscopic parameters related to interactions. However, the avalanche is difficult to be described via these semi-classical models, as the impact-ionization process itself depends at nearly atomic level on the generated fields, which in turn depend on the generated charge. In particular, the ionization coefficients for electrons and holes can be calculated for biasing up to the breakdown voltage, but for higher polarizations the charges generated by impact ionization modify locally the electric field, so that the coefficients get stabilized and no further progress can be calculated for any overvoltage. Solving this situation needs further advances in models.

Consequently, proprietary software needs to be developed and applied to properly describe the avalanche, and then, well established electrical models will help describing the device inside the whole detector system. The complete simulation process is complex, but necessary to understand the device behaviour in operating systems as for characterization.

3.2 I-V characteristics

The breakdown voltage is evidenced by a sudden current increase of several orders of magnitude in I-V characteristics for reverse bias. It is strongly technology dependent, because it is related to concentration and depth of the doping profiles, and can even be sensitive to the particular run and chip, as illustrated in fig. 7 (Arbat et al., 2010b). The analysis of the temperature effect on the breakdown voltage indicates that both parameters increase together. All these variations can be managed with adequate readout electronics.

3.3 Dark count rate

The dark-count rate increases exponentially with biasing, as shown in figure 8, and approximately linearly with the detector area. However, important differences can be found depending on the technology used. In particular, the trench-based structures present much more dark counts than the structure based on an n well. This suggests that the fabrication of these trenches introduces traps in their walls, increasing the pulse-generation probability (Hamamoto, 1991). Consequently, trench-based technologies should be treated with care when a low noise rate is required. The influence of other factors such as degradation due to radiation is still to be studied.

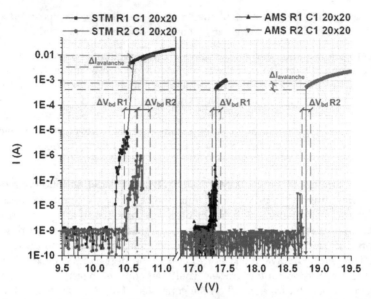

Fig. 7. I-V characteristics of 20μmx20μm APDs made using STM (left) and HV-AMS (right) technologies, for two different runs (R1 and R2) each.

An analysis with the temperature allows observing that the number of dark-count pulses increases exponentially with the temperature, as expected due to thermal generation of electron-hole pairs. The dynamic-range analysis of these types of structures gives sensibilities with similar behaviour independently from the technology, incrementing the counts linearly with illumination, until saturation is reached. On the other hand, increasing biasing raises the sensor sensibility.

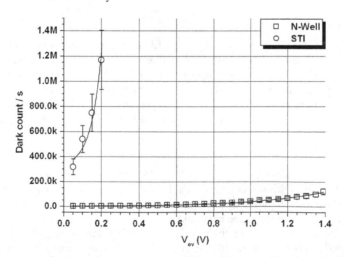

Fig. 8. Dark-count rate for two detector structures based on n-well and on shallow-trench isolation, STI, respectively, vs. operation voltage, V_{op}.

3.4 Time response and afterpulsing

Direct displaying of the voltage measurements vs. time indicates the afterpulsing behaviour of a sensor. Figure 9 shows the noise counts shortly after the pulse generation. Although the probes capacitance introduces some deformation of the signal, this type of graph also allows analyzing in some detail the time response of the sensor. For different technologies there are clear different timing behaviours, which can be summarized in a longer quenching time for n-well structures as compared to trench based ones, what favours the probability of charge trapping and increases afterpulsing. When the quenching and readout circuits are introduced into the ASIC, this quenching time can be reduced, as well as the probability of afterpulsing. Another noticeable remark is the longer dead time for n-well than for trench structures, which should be reduced by implanting active-recharge systems.

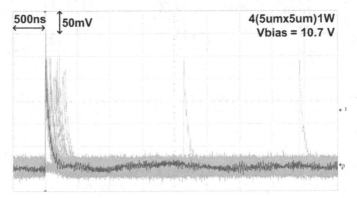

Fig. 9. Afterpulsing and noise visible in direct displaying vs. time.

3.5 Sensitive areas and cross-talk in APD arrays

A novel method has been recently reported to characterize APD-arrays response with high space resolution, whose results suggest that the fill factor can be drastically improved via good control of the doping profiles (Vilà et al., 2011). This research represents a novel effort to use APDs to construct an active pixel detector for charged particles in colliders. The key issue consist on using a focused ion beam/scanning electron microscopy, FIB/SEM setup to precisely characterize the sensitive areas and the quantum efficiency. The beams provided by FIB/SEM setup are nanometric, in contrast to usual test beams that lighten the detectors with lasers providing spots of several tens of microns -comparable to detector dimensions.

A FIB/SEM setup allows a beam to be accelerated, collimated into a nanometric spot and focused with nanometric precision onto one pixel, as well as to perform a detection experiment without trigger setup, thanks to the controlled amount of particles injected per measuring time. The beam is scanned from a pixel to its neighbour one, as shown in figure 10-left, and the agreement between the mathematical description of the detection events and the measurement corroborates both. The measured detection profiles are as shown in figure 10-right, where different applied bias are represented. The position and separation between adjacent pixels is marked onto the profiles taking onto account their distance on the layout, the width of the guard rings and their lateral diffusion.

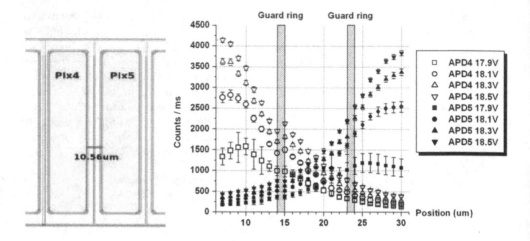

Fig. 10. Layout of the pixels used for sensitive-area and crosstalk demonstration (left). Detection frequencies measured by scanning a 30kV electron beam onto these two neighbouring pixels (right).

It is important to remark that this new characterization is only useful with a detailed functional simulation. The qualitative agreement between Monte Carlo calculations and experimental detection frequencies under a low-flow electron beam (Vilà et al., 2011) avails the correctness of both and suggests good efficiency for the APD prototypes. The nanometric dimension of the beam, as well as its accurate positioning onto the APD array, allows a detailed study of the dead areas around pixels, where the high detection rate observed suggests a noticeable efficiency of the guard ring as detector. This situation could improve drastically the fill factor of the devices, which could virtually arrive to 100%.

4. Circuitry involved in the G-APD operation

The improvement of the sensor performances has been strongly influenced by the development of the related electronics. In the case of dedicated technologies, the quenching electronics is implemented externally to the sensor. In the case of monolithic technologies, the quenching and any additional signal processing can be done close to the sensor, allowing an increase of functionality while reducing area, consumption and power requirements. In this section, basic circuits will be revised, starting from some SPICE models for describing the behaviour of the APDs.

4.1 Models for APD detectors

In order to develop a complete electronic system involving the sensor, an electrical model will be necessary for it. Few models are already available to reproduce partially the behaviour of the APD, but they do not offer a complete vision of the sensor. Recent developments are summarized in this section.

Since the initial investigations of the p-n junction characteristics, developed after the Shockley prediction (Shockley, 1961) and confirmed on the early '60s (Goetzberg et al., 1963), the multiplication behaviour of such structures working in Geiger mode generated a great expectation. The first electrical model was developed as a consequence of these studies (Haitz, 1965) and has been largely used. The model contemplates the parasitic capacitances of the device and reproduces the junction behaviour when a photon is detected. Recently, a second model was presented, using programmable voltage switches (Dalla Mora et al., 2007). In this model, a voltage programmable source is used to reproduce the different behaviours of the APD current depending on the bias. It also contemplates the self-sustaining process reached in APDs when the current rises to a certain threshold. An alternative to this model includes a voltage-controlled switch to reproduce both the static and dynamic behaviours (Mita et al., 2008). The switch is defined in Verilog-A language, and the model also deals with the dependence of the device capacitance on the reverse bias. A newer modification includes the current-voltge behaviour of a diode in the whole polarization range. It adds forward behaviour and the second breakdown (Zappa et al., 2009). The model is presented for both SPICE and Virtuoso Spectre simulators. All these three fundamental models are presented in figure 11.

However, all the models presented so far lack to consider uncorrelated (dark counts) and pulse-correlated (afterpulsing) noises. The electronics used after the sensor has to be designed considering that these false hits must be stored for postprocessing. The first model including diode behaviour for all the polarization range and noise reproduction has been presented recently (Arbat et al., 2010b). Its validity is demonstrated by adjusting the measurements obtained after the characterization of two different APD structures fabricated with a standard 130nm CMOS processing.

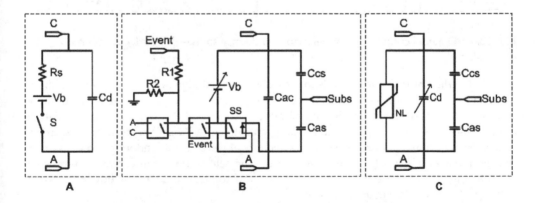

Fig. 11. Models for APDs operation: A) (Haitz et al., 1965), B) (Dalla Mora et al., 2007) and C) (Mita et al., 2008).

This new model uses a source with a noise pattern, together with other active and passive components from Spectre (Cadence™), to reproduce the noise behaviour in the time domain. (Arbat, 2010b). The noise transitory analysis available from the Cadence™ software since version 2008 (Cadence, 2008) allows the possibility of reproducing the actual sensor behaviour. Including the noise behaviour in the model it is possible to obtain information about the probability of information losses due to the dead time related to the noise generation. It also allows studying the strategies that can be used for avoiding the noise interference into the signal, such as gated-mode acquisition or active quenching and recharge circuits. Finally, the analysis of architectures for array readout considering the generated noise, indistinguishable from the signal, is also possible.

The model is represented in figure 12. It contains the elements presented in the previous models, such as anode-cathode capacitor, C_{ac}, the anode-bulk capacitor, C_{ab}, and the cathode-bulk capacitor, C_{cb}. However, the different I-V regions are reproduced with three different groups of resistor, voltage source and voltage controlled switch. The first branch models the forward behaviour of the diode, V_f, allowing the current flow when the forward bias is higher than the bandgap. The second branch represents the second breakdown (Vb2) related to the breakdown of the lateral sensor junctions.

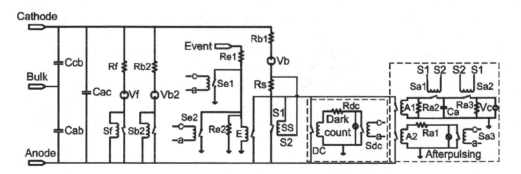

Fig. 12. APD model containing I-V behaviour and noise sources for dark counts and afterpulsing.

Finally, the third branch corresponds to the Geiger-mode region (V_b), and has been modified from the previous models to include noise effects. It offers four possible ways to allow conductivity between both terminals: photon/particle arrival (E), self-sustaining avalanche (SS), dark-count event (DC) and afterpulsing generation (A1-A2). To reproduce the spurious nature of dark-count pulses, the source of the model generates a random noise signal with the frequencies obtained during the sensor characterization. Similarly, to reproduce afterpulsing events, the sensing resistor Rs is used to sense the current flowing between anode and cathode during an avalanche.

The novelty of this model is to add these sources of noise to an electrical model of the sensor, and its validity has been tested by comparison with experimental measurements obtained after characterization of two APD structures fabricated with a standard 130nm CMOS process. Figures 13 A, B and C summarize the obtained results, demonstrating the validity of this newest model for APDs.

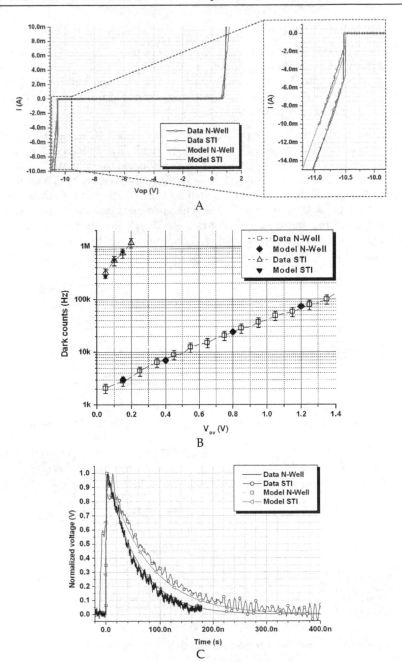

Fig. 13. A) I-V measurements and simulation using the models for an N-well and a STI structures. B) Dark counts measured for the N-well and the STI structures with a 10 kΩ quenching resistance. C) Time response of the two structures and the models represented with a normalized voltage and a quenching resistance of 10 kΩ.

4.2 Quenching and recharge circuits

The dead time, i.e. the total time between the start of a pulse and the resetting of the bias voltage after quenching and recharge, limits the device speed. Many different circuits have been developed to minimize this time, which can be divided into active, passive or combination of both. In all cases, the avalanche reading is done by converting analogue signal to digital by means of a comparator.

As the simplest approximation, a passive quenching circuit (Cova et al., 1996) can be obtained by connecting a high-valued quenching resistance, R_q, in series to the APD (fig. 14). As the avalanche current starts to flow, the resistance forces the operating voltage, V_{op}, to drop down to the breakdown voltage, V_{bd}, quenching the avalanche. This process is not instantaneous because of the parasitic capacitance of the diode, but follows an exponential decay determined by $R_q C_{APD}$ and the avalanche current. Also the circuit recovery after quenching is exponential, but slower owing to the recharge current is much lower than the avalanche current. The sum of the quenching, T_q, and recharge, T_r, times corresponds to the dead time, T_{dead}, which in this case is quite long.

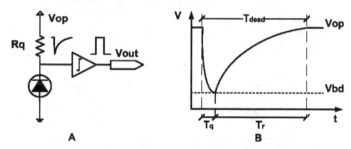

Fig. 14. A) Both passive quenching and recharge circuits. B) Sketch of the timing behaviour.

The next development of the passive quenching for integrated APDs corresponds to the use of a saturated MOS transistor as quenching resistance (figure 15). During the recharge period, the transistor works as a variable resistor, thanks to the change in the biasing.

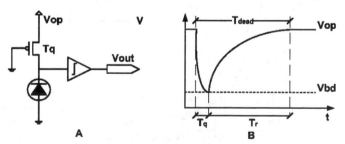

Fig. 15. A) Passive quenching by a transistor and detecting circuits. B) Sketch of the timing behaviour.

As the main contribution to the dead time is due to the recharge stage, a first improvement can be obtained by using active recharge. A simple circuit to provide it consists of an additional MOSFET with low capacitance as switch in parallel to the quenching resistor.

The circuit in figure 16 includes a 50Ω resistor R_s to match impedances with the measuring circuit. When the avalanche is detected, the comparator generates signal, the switch is closed and the sensor recovers the operating voltage, V_{op}. When the process is finished, the switch is opened again and the detector becomes ready for a new detection. This implementation reduces the dead time, but it can increase the afterpulsing noise, as the traps filled during the avalanche may not be released.

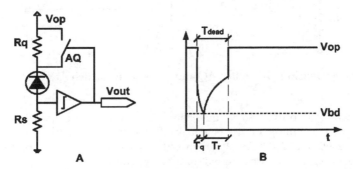

Fig. 16. A) Passive quenching and active recharge by a switch AQ. B) Sketch of the timing behaviour.

Using an active quenching can reduce afterpulsing noise, as the probability for a charge to be trapped is proportional to the duration of the avalanche event. A simple way to generate an active quench is to connect a switch between the APD and a voltage source below the breakdown (figure 17). When the avalanche is detected, the switch is closed and the sensor is quickly biased to this low voltage, producing fast avalanche quenching and limiting the charge through the detector.

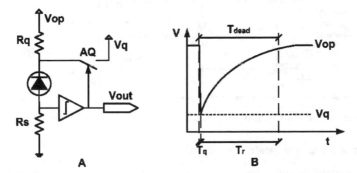

Fig. 17. A) Active quenching with passive recharge. B) Sketch of the timing behaviour.

The most complete approach to improve timing consists of both active quenching and recharge (figure 18). When an avalanche pulse is sensed, the comparator activates the voltage driver, which switches AQ and lowers the biasing down to the breakdown voltage. After a hold-off time (not represented in the figure), the bias voltage is switched back to the operating level by using AR. This circuit reduces the sensor dead time allowing to reach rates of million counts per second while maintaining the high resolution in photon timing.

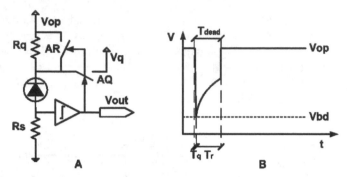

Fig. 18. A) Active quenching by switch AQ and recharge by switch AR. B) Sketch of the timing behaviour.

The development of improved active quenching and recharge circuits has been important. Integrated circuits including improvements of this simple approach have been reported (Zappa et al., 2000 and 2002) involving mixed passive-active circuit configurations that combined integrated circuit with external components (figure 19).

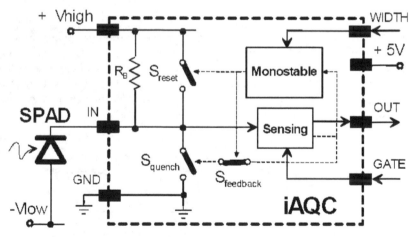

Fig. 19. Integrated active quenching and recharge structure (Zappa et al., 2002).

4.3 Operation modes and readout methods

APDs can be always reverse-biased at a fixed voltage above V_{BD}, so that they are always ready to detect (free running). However, in those applications where the signal arrival time is known, the sensor can also be activated only in certain periods (gated acquisition). In contrast with the free-running operation mode, in the gated acquisition mode the reverse bias voltage swings from over to under V_{BD} to periodically enable and disable the photodiode. In this way, the sensor is kept active only for short periods of time that can be synchronized with the expected signal arrival. Consequently, the probability to detect dark counts interfering with the signal-triggered ones is linearly reduced with the width of the active period of the sensor without missing any photon counts, and hence the signal/noise ratio is improved (Vilella et al., 2011a and b). In addition, long enough non-active periods, longer than the lifetime of the trapping levels, allow to completely eliminate the afterpulsing probability. The basic circuits for running in gated mode can be seen in figure 20.

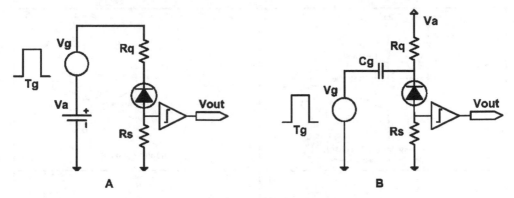

Fig. 20. Circuits for gated-mode operation. A) DC coupling and B) AC coupling.

Recently, a new prototype having the front-end electronics for gated acquisition monolithically integrated with the G-APD has been demonstrated (Vilella et al., 2011a and b). The pixel detector, based on a gated G-APD with passive quenching and active recharge (through RST transistor), is shown in figure 21. The operation mode, together with suitable readout electronics that allows low reverse overvoltage, reduces both the dark-count rate and probability, and eliminates afterpulsing at the pixel. Also, gating with shorter observation periods allows a considerable increase in the sensor dynamic range.

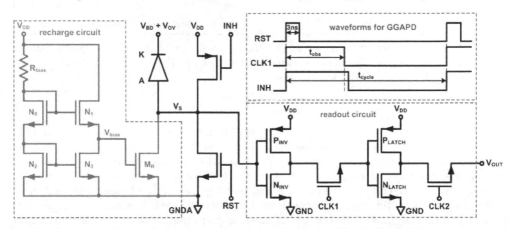

Fig. 21. Schematic diagram of the pixel detector integrated with gated-acquisition electronics, with the waveforms of the gated acquisition, in the inset up right (Vilella et al., 2011b).

Finally, several readout circuits for CMOS G-APDs allowing low-noise operation have been also reported (figure 22), to complete the previous sensor with gated acquisition (Vilella et al., 2010 and 2011b). By using a front-end circuit based on a floating ground, avalanche detections were possible for very low overvoltage, which also reduced the dark-count rate. The characterization of this pixel showed that it could provide acceptable performance even for high-speed applications.

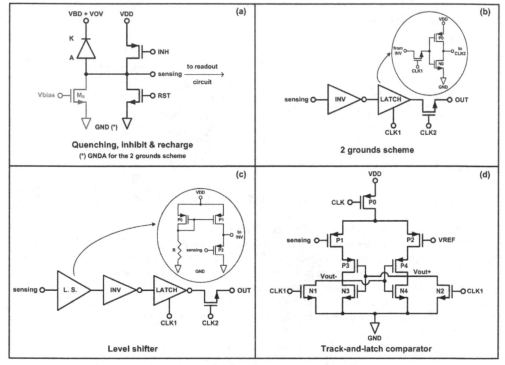

Fig. 22. A) Generic schematic diagram of the structure with gated acquisition. B) Readout circuit based on a 2-grounds scheme, C) on a level shifter and D) in a track-and-latch comparator (Vilella et al., 2011b).

4.4 Readout options for G-APDs arrays

Finally, G-APDs are usually used in large area applications that require their organization in sensor arrays or cameras. The reading options of a camera are mainly related to its final application: for imaging every pixel must be read, while for an even-driven information, such as the position of a particle arrival, only the information of the hit pixels is needed. Consequently, a detector array can be read sequentially, i.e. all the elements are read in an ordered way, or event-driven, what means reading only the pixels in which some signal has been detected. The sequential readout can also be performed with one or several output lines (figure 23 A and B), being possible to adjust the data-transfer speed to the application-required speed. In this case, every pixel contains the basic quenching and recharging circuits together with the elements for selecting row and column (Niclass & Charbon, 2005). On the contrary, in event-driven readout only the pixels that have received a pulse are read (figure 23 C), and this is optimal when only a small part of the array receives information. In this case, the electronics becomes more complicated than for sequential readout, but the timing is noticeably reduced (Niclass et al., 2006a).

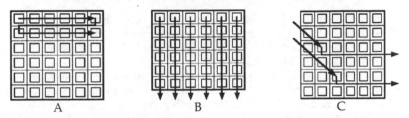

Fig. 23. Sequential readout for a detector array, with only one (A) or several (B) outputs. In C, event-driven readout mechanism.

After reading, signal must be processed, and, for some applications, the implementation of a part of this processing in the pixel can be very interesting (figure 24). In this case, the pixel-contained electronics becomes complicated, and readout of the whole camera after the complete measurement is required. A simplest circuit could be a pulse counter able to provide the image gray levels (Tisa et al., 2008 and 2009).

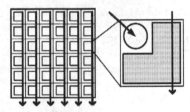

Fig. 24. Electronic processing inside the pixel.

5. Summary and conclusions

A general overview of the main aspects related to Geiger-mode avalanche photodiodes has been exposed in this chapter. Those sensors have been largely used for single photon detection, and nowadays are also used for X- and gamma-rays detection and high-energy particles. After the revision of the main concepts related to their operation, dedicated and standard technologies have been considered from the historical review of their respective evolution. The different aspects corresponding to the different types of sensors, the related circuits and the measurement and definition of the principal figures of merit of the sensors have been presented. Recent developments of the scientific community related to the use of G-APDs have also been exposed here.

6. Acknowledgment

This work has been partially supported by the Spanish projects "Desarrollo de nuevas tecnologías en aceleradores y detectores para los futuros colisionadores de física de partículas", coded FPA2008-05979-C04-02 and "Desarrollo de nuevos detectores para los futuros colisionadores en Física de Partículas", coded FPA2010-21549-C04-01.

7. References

Alexander, S.B. (1997). *Optical communication receiver design (SPIE tutorial texts in optical engineering Vol. TT22)*, SPIE publications, IEE Telecommunications Series Vol. 37, ISBN 0-8194-2023-9, Bellingham WA

Arbat, A., Trenado, J., Gascon, D., Vilà, A., Comerma, A., Garrido, L. & Dieguez, A. (2010a). High voltage vs. high integration: a comparison between CMOS technologies for SPAD cameras. *Proc. of SPIE Optics and Photonics 2010*, Vol. 7780, pp. 77801G, ISSN 0277-786X, San Diego CA, August 2010

Arbat, A. (2010b). Towards a forward tracker detector based on Geiger mode avalanche photodiodes for future linear colliders. PhD dissertation, December 2010

Assefa, S., Xia, F. & Vlasov, Y.A. (2010). Reinventing germanium avalanche photodetector for nanophotonic on-chip optical interconnects. *Nature*, Vol. 464, (March 2010), pp. 80-85, ISSN 0028-0836

Cadence. (2008). Application notes on direct time-domain noise analysis using virtuoso spectre, version 1.0

Charbon, E. (2007). Will avalanche photodiode arrays ever reach 1 megapixel?. *Proc. 2007 International Image Sensor Workshop*, pp. 246-249, Ogunquit ME, 2007

Charbon, E. (2008). Towards large scale CMOS single-photon detector arrays for lab-on-chip applications. *Journal of Physics D: Applied Physics* Vol. 41, (2008), pp. 094010-1:9, ISSN 0022-3727

Cova, S., Ghioni, M., Lacaita, A., Samori, C. & Zappa, F. (1996). Avalanche Photodiodes and Quenching Circuits For Single-Photon Detection. *Applied Optics*, Vol. 35, No. 12 (April 1996), pp. 1956-1976, ISSN 1943-8206

Cova, S., Ghioni, M., Lotito, A., Rech, I. & Zappa, F. (2004). Evolution and prospects for single-photon avalanche diodes and quenching circuits. *Journal of Modern Optics*, Vol. 51, No. 9-10 (June-July 2004), pp. 1267-1288, ISSN 0950–0340

Dalla Mora, A., Tosi, A., Tisa, S. & Zappa, F. (2007). Single-Photon Avalanche Diode Model for Circuit Simulations. *IEEE Photonics Technology Letters*, Vol. 19, No. 23 (December 2007), pp. 1922-1924, ISSN 1041-1135

Dandin, M., Nelson, N., Saveliev, V., Ji, H., Abshire, P. & Weinberg, I. (2007). Single photon avalanche detectors in standard CMOS. *IEEE Sensors '07*, ISSN 1930-0395, pp. 585-588, Atlanta, October 2007

Dautet, H., Deschamps, P., Dion, B., MacGregor, A.D., MacSween, D., McIntyre, R.J., Trottier, C. & Webb, P.P. (1993). Photon counting techniques with silicon avalanche photodiodes. *Appl. Opt.*, Vol. 32, No. 21 (1993), pp. 3894-3900, ISSN 2155-3165

Dussault, D. & Hoess, P. (2004). Noise Performance Comparison of ICCD with CCD and EMCCD Cameras, *Proceedings of SPIE*, Vol. 5563, pp. 195-204, ISBN 0-8194-5501-6, Denver, August 2004

EG&G (1996) Canada, Vaudreuil, P.Q., Canada, SPCM-AQ Series Data Sheet ED-0043/11/96, 1996

Faramarzpour, N., Deen, M.J., Shirani, S. & Fang, Q. (2008). Fully integrated single-photon avalanche diode detector in standard CMOS 0.18 μm technology. *IEEE Transactions on electron devices*, Vol. 55, No. 3 (March 2008), pp. 760-767, ISSN 0018-9383

Finkelstein, H., Hsu, M.J. & Esener, S.C. (2006a). STI-Bounded single-photon avalanche diode in a deep-submicrometer CMOS technology. *IEEE Electron Device Letters*, Vol. 27, No. 11 (November 2006), pp. 887-889, ISSN 0741-3106

Finkelstein, H., Hsu, M.J. & Esener, S. (2006b). An ultrafast Geiger-mode single photon avalanche diode in 0.18μm CMOS technology. *Advanced photon counting techniques, Proc. of SPIE*, Vol. 6372, ISBN 9780819464705, pp. 63720W-1:10, Boston MA, October 2006

Gersbach, M., Niclass, C., Charbon, E., Richardson, J., Henderson, R. & GRANT, L. (2008). A single photon detector implemented in a 130 nm CMOS imaging process. *Proc. of the 38th European Solid-State Device Research Conference, ESDERC 2008*, pp. 270-273, ISSN 1930-8876, Edinburgh, September 2008

Gersbach, M., Richardson, J., Mazaleyrat, E., Hardillier, S., Niclass, C., Henderson, R., Grant, L. & Charbon, E. (2009). A low-noise single-photon detector implemented in a 130 nm CMOS imaging process. *Solid-State Electronics*, Vol. 53, (May 2009), pp 803-808, ISSN 0038-1101

Ghioni, M., Cova, S., Lacaita, A. & Ripamonti, G. (1988). New silicon epitaxial avalanche diode for single-photon timing at room temperature, *Electron. Lett.*, Vol. 24, No. 24 (Nov. 1988), pp 1476-1477, ISSN 0013-5194

Ghioni, M. & Ripamonti, G. (1991). Improving the performance of commercially available Geiger-mode avalanche photodiodes. *Rev. Sci. Instrum.*, Vol. 62, No. 1 (January 1991), pp. 163-167, ISSN 0034-6748

Ghioni, M., Gulinatti, A., Rech, I. & Cova, S. (2006). Recent advances in silicon single photon avalanche diodes and their applications. *Annual Meeting of the IEEE Lasers and Electro-Optics Society, LEOS 2006*, ISBN 0-7803-9555-7, pp. 719-720, Montreal, October 2006

Ghioni, M., Gulinatti, A., Rech, I., Zappa, F. & Cova, S. (2007). Progress in silicon single-photon avalanche diodes. *IEEE Journal of selected topics in quantum electronics*, Vol. 13, No. 4 (July-August 2007), pp. 852-862, ISSN 1077-260X

Goetzberg, A., McDonald, B., Haitz, R. & Scarlett, R. (1963). Avalanche effects in silicon p-n junctions. II. Structurally perfect junctions. *Journal of App. Phys.*, Vol. 34, (1963), pp. 1591-1600 ISSN 0021-8979

Graugés, E., Comerma, A., Garrido, L., Gascón, D., Trenado, J., Diéguez, A., Vilà, A., Arbat, A., Freixas, L., Hidalgo, S., Fernández, P., Flores, D. & Lozano, M. (2010). Study of Geiger avalanche Photo-diodes (Galds) applications to pixel Trucking detectors. *Nucl. Inst. Met. Phys. Res. A*, Vol. 617, (2010), pp. 541-542, ISSN 0168-9002

Guerrieri, F., Tisa, S. & Zappa, F. (2009). Fast Single-Photon imager acquires 1024 pixels at 100 kframe/s. *Proceedings of SPIE*, Vol. 7249, pp. 72490U-1 – 72490U-11, ISSN 0277-786X, San José CA, January 2009

Gulinatti, A., Maccagnani, P., Rech, I., Ghioni, M. & Cova, S. (2005). 35 ps time resolution at room temperature with large area single photon avalanche diodes. *Electronic letters*, Vol. 41, No. 5 (March 2005), pp. 272-274, ISSN 0013-5194

Haitz, R. (1964). Model for the electrical behavior of a microplasma. J. Appl. Phys., Vol. 35, (1964), pp. 1370-1376 ISSN 0021-8979

Haitz, R. (1965). Mechanisms contributing to the noise pulse rate of avalanche diodes. J. Appl. Phys., Vol. 36, (1965), pp. 3123-3131, ISSN 0021-8979

Hamamoto, T. (1991). Sidewall damage in a silicon substrate caused by trench etching. *Applied physics letters*, Vol. 58, No. 25 (June 1991), pp. 2942-2944, ISSN 0003-6951

Jackson, J.C., Morrison, A.P. & Lane, B. (2000). Characterization of large area SPAD detectors operated in avalanche photodiode mode. *IEEE 13th Annual Meeting Laser and Electro-Optics Society, LEOS 2000*, Vol. 1, pp. 17-18, ISBN 0-7803-5947-X, Rio Grande, November 2000

Jackson, J.C., Morrison, A.P., Hurley, P., Harrell, W.R., Damjanovic, D., Lane, B. & Mathewson, A. (2001). Process monitoring and defect characterization of single photon avalanche diodes. *Proc. of the 2001 International Conference on Microeletronic Test Structures*, Vol. 14, pp. 165-170, ISBN 0-7803-6511-9, Kobe, March 2001

Jackson, J.C., Morrison, A.P., Phelan, D. & Mathewson, A, (2002). A novel silicon Geiger-mode avalanche photodiode. *Digest. International Electron Devices Meeting, IEDM'02*, ISBN 0-7803-7462-2, pp. 797-800, San Francisco CA, December 2002

Karami, M.A., Gersbach, M., Yoon, H.J. & Charbon, E. (2010). A new single-photon avalanche diode in 90nm standard CMOS technology. *Optics Express*, Vol. 18, No. 21 (October 2010), pp. 22158-22166, ISSN 0146-9592

Kindt, W.J. (1994). A novel avalanche photodiode array. *Nuclear Science Symposium and Medical Imaging Conference*, 1994 IEEE Conference Record, Vol. 1, (Oct-Nov 1994), pp. 164-167 ISBN 0-7803-2544-3, Norfolk, VA, Oct-Nov 1994

Lacaita, A., Cova, S. & Ghioni, M. (1988). Four-hundred-picosecond single-photon timing with commercially available avalanche photodiodes. *Rev. Sci. Instrum.*, Vol. 59, no. 7 (July 1988), pp. 1115-1121, ISSN 0034-6748

Lacaita, A., Ghioni, M. & Cova, S. (1989) Double epitaxy improves single-photon avalanche diode performance. *Electron. Lett.*, Vol. 25, No. 13 (June 1989), pp. 841-843, ISSN 0013-5194

Lacaita, A.L., Zappa, F., Bibliardi, S. & Manfredi, M. (1993). On the bremsstrahlung origin of hot-carrier-induced photons in silicon devices. *IEEE Transactions on Electron Devices*, Vol. 40, No. 3 (March 1993), pp. 577-582, ISSN 0018-9383

McIntyre, R. (1972). The distribution of gains in uniformly multiplying avalanche photodiodes: Theory, *IEEE trans. Electron Devices*, Vol. 19, No. 6 (June 1972), pp 703-713, ISSN 0018-9383

McIntyre, R. (1990). Silicon avalanche photodiode with low multiplication noise. *US patent* n° 4,972,242"

McIntyre, R. & Webb, P. (1996). Low-noise, reach-through, avalanche photodiodes. *US Patent* n° 5,583,352"

McKay, K.G. (1954). Avalanche breakdown in silicon. *Phys. Rev.*, Vol. 94, No. 4 (1954), pp. 877–884, ISSN 1943-2879

Mita, R., Palumbo G. & Fallica, P.G. (2008). Accurate model for single-photon avalanche diodes. *IET circuits devices systems*, Vol. 2, No. 2 (April 2008), pp. 207–212, ISSN 1751-858X

Mosconi, D., Stoppa, D., Pancheri, L., Gonzo, L. & Simoni, A. (2006). CMOS single-photon avalanche diode array for time-resolved fluorescence detection. *Proc. of the 32nd European Solid-State Circuits Conference, ESSCIRC'06*, ISSN 1930-8833, pp. 564-567, Montreux, September 2006

Niclass, C. & Charbon, E. (2005). A single photon detector array with 64x64 resolution and millimetric depth accuracy for 3D imaging. *IEEE International Solid-State Circuits Conference, ISSCC 2005*, pp. 364-366, ISSN 0193-6530, San Francisco, February 2005

Niclass, C., Sergio, M. & Charbon, E. (2006a). A CMOS 64x48 single photon avalanche diode array with event-driven readout. *Proc. of the 32th European SolidState Circuits Conference, SSCIRC'06*, pp. 556-559, ISBN 1-4244-0303-4, Montreux, September 2006

Niclass, C., Sergio, M. & Charbon, E. (2006b). A single photon avalanche diode array fabricated in 0.35μm CMOS and based on an event-driven readout for TCSPC experiments. *Proc. of SPIE, Advanced Photon Counting Techniques*, Vol. 6372, ISBN 9780819464705, pp. 63720S-1:12, Boston MA, October 2006

Niclass, C., Gersbach, M., Henderson, R., Grant, L. & Charbon, E. (2007a). A single photon avalanche diode implemented in 130 nm CMOS technology, *IEEE Journal of selected topics in quantum electronics*, Vol. 13, No. 4 (July/August 2007), pp. 863-869, ISSN 1077-260X

Niclass, N., Gersbach, M., Henderson, R.K., Grant, L. & Charbon, E. (2007b). A 130 nm CMOS single photon avalanche diode. *Proceedings of SPIE*, Vol. 6766 (2007), pp. 676606, ISSN 0277-786X, Boston, September 2007

Niclass, C., Favi, C., Kluter, T., Monnier, F. & Charbon, E. (2009). Single-photon synchronous detection. *IEEE Journal of Solid-State Circuits*, Vol. 44, No. 7 (July 2009), pp. 1977-1989, ISSN 0018-9200

Pancheri, L. & Stoppa, D. (2007). Low-noise CMOS single-photon avalanche diodes with 32 ns dead time. *37th European Solid State Device Research Conference, ESSDERC 2007*, ISSN 1930-8876, pp. 362-365, Munich, September 2007

Renker, D. (2004). Photosensors. *Nucl. Inst. and Meth. in Phys. Res. A*, Vol. 527, No. 1-2, (July 2004), pp. 15-20, ISSN 0168-9002

Richardson, J.A., Grant, L.A. & Henderson, R.K. (2009). Low dark count single-photon avalanche diode structure compatible with standard nanometer scale CMOS technology. *IEEE Photonics Technology Letters*, Vol. 21, No. 14 (July 2009), pp. 1020-1022, ISSN 1041-1135

Rochas, A., Pauchard, A.R., Besse, P.A., Pantic, D., Prijic, Z. & Popovic, R.S. (2002). Low-Noise silicon avalanche photodiodes fabricated in conventional CMOS technologies. *IEEE Transactions on Electron Devices*, Vol. 49, No. 3 (March 2002), pp. 387-394, ISSN 0018-9383

Rochas, A., Gani, M., Furrer, B., Besse, P.A., Popovic, R.S., Riborby, G. & Gisin, N. (2003a). Single photon detector fabricated in a complementary metal-oxide-semiconductor high-voltage technology. *Rev. Sci. Instrum.*, Vol. 74, No. 7 (July 2003), pp. 3263-3270, ISSN 0034-6748

Rochas, A., Gösch, M., Serov, A., Besse, P.A., Popovic, R.S., Lasser, T. & Rigler, R. (2003b). First Fully Integrated 2-D Array of Single-Photon Detectors in Standard CMOS Technology. *IEEE Photonics Technology Letters*, Vol. 15, No. 7 (July 2003), pp. 963-965, ISSN 1041-1135

Sergio, M., Niclass, C. & Charbon, E. (2007). A 128x2 CMOS single-photon streak camera with timing-preserving latchless pipeline readout. *IEEE International Solid-State Circuits Conference, ISSCC 2007*, pp. 394-396, ISSN 0193-6530, San Francisco, February 2007

Shockley, W. & Read, W.T. (1952). Statistics of the recombinations of holes and electrons. *Phys. Rev.*, Vol. 87, No. 5 (September 1952), pp. 835-842

Shockley, W. (1961). Problems related to p-n junctions in silicon. *Solid-State Electron*. Vol. 2, No. 1 (January 1961), pp. 35-60, ISSN 0038-1101

Spinelli, A., Davis, L.M. & Dautet, H. (1996). Actively quenched single photon avalanche diode for high repetition rate time-gated photon counting. *Rev. Sci. Instrum.*, Vol. 67, No. 1 (January 1996), pp. 55-61, ISSN 0034-6748

Spinelli, A., Ghioni, M.A., Cova, S.D. & Davis, L.M. (1998). Avalanche detector with ultraclean response for time-resolved photon counting. *IEEE Journal of Quantum Electronics*, Vol. 34, No. 5 (May 1998), pp. 817-821, ISSN 0018-9197

Stapels, C.J., Lawrence, W.G., Augustine, F.L. & Christian, J.F. (2006). Characterization of a CMOS Geiger photodiode pixel. *IEEE Transactions on electron devices*, Vol. 53, No. 4 (April 2006), pp. 631-635, ISSN 0018-9383

Stapels, C.J., Squillante, M.R., Lawrence, W.G., Augustin, F.L. & Christian, J.F. (2007). CMOS-based avalanche photodiodes for direct particle detection. *Nucl. Inst. and Meth. In Phys. Res. A*, Vol. 579, No. 1 (2007), pp. 94-98, ISSN 0168-9002

Stoppa, D., Pancheri, L., Scandiuzzo, M., Simoni, A., Viarani, L. & Dalla Betta, G.F. (2005). A CMOS sensor based on signle photon avalanche diode for distance measurement applications. *Proc. of the IEEE. Instrumentation and measurement technology conference*, Vol. 2, ISBN 0-7803-8879-8, pp. 1162-1165, Ottawa, May 2005

Stoppa, D., Pancheri, L., Scandiuzzo, M., Gonzo, L., Dalla Betta, G.F. & Simoni, A. (2007). A CMOS 3-D imager based on single photon avalanche diode. *IEEE Transactions on Circuits and Systems - I*, Vol. 54, No. 1 (January 2007), pp. 4–12, ISSN 1057-7122

Tisa, S., Guerrieri, F. & Zappa, F. (2008). Variable-load quenching circuit for single-photon avalanche diodes. *Optics express*, Vol. 16, No. 3 (February 2008), pp. 2232-2244, ISSN 0146-9592

Tisa, S., Guerrieri, F. & Zappa, F. (2009). Monolithic array of 32 SPAD pixel for single-photon imaging at high frame rates. (2009). *Nucl. Instr. Meth. Phys. Res. A:*

Accelerators, Spectrometers, Detectors and Associated Equipment, Vol. 610, No. 1 (October 2009), pp. 24-27, ISSN 0168-9002

Vilà, A., Trenado, J., Arbat, A., Comerma, A., Gascon, D., Garrido, Ll. & Dieguez, A. (2011). Characterization and simulation of avalanche photodiodes for next-generation colliders. *Sensors and Actuators A: Physical*, (2011) doi:10.1016/j.sna.2011.05.011

Vilella, E., Arbat, A., Comerma, A., Trenado, J., Alonso, O., Gascon, D., Vilà, A., Garrido, L & Dieguez, A. (2010). Readout electronics for low dark count pixel detectors based on Geiger mode avalanche photodiodes fabricated in conventional CMOS Technologies for future linear colliders. *Nucl. Instr. Meth. Phys. Res. A*, Vol. 650, (2010), pp. 120-124, ISSN 0168-9002

Vilella, E., Comerma, A., Alonso, O. & Dieguez, A. (2011a). Low-noise pixel detectors based on gated Geiger mode avalanche photodiodes. *Electronics Letters*, Vol. 47, No. 6 (March 2011), ISSN 0013-5194

Vilella E., Arbat, A., Comerma, A., Trenado, J., Alonso, O., Gascon, D., Vilà, A., Garrido, L. & Dieguez, A. (2011b). Readout electronics for low dark count Geiger mode avalanche photodiodes fabricated in conventional HV-CMOS technologies for future linear colliders. *Topical Workshop on Electronics for Particle Physics 2010*, pp. C01015, Aachen, September 2011

Webb, P., McIntyre, R. & Conradi, J. (1974). Properties of avalanche photodiodes. *RCA Rev.*, Vol. 35, (June 1974), pp. 234-278, ISSN 0033-6831

Xiao, Z., Pantic, D. & Popovic, R.S. (2007). A new single photon avalanche diode in CMOS high-voltage technology. *International solid-state sensors, actuators and microsystems conference, Transducers 2007*, ISBN 1-4244-0842-3, pp. 1365-1368, Lyon, June 2007

Zappa, F., Ghioni, M., Cova, S., Varisco, L., Sinnis, B., Morrison, A. & Mathewson, A. (1997). Integrated array of avalanche photodiodes for single-photon counting. *Proc. 27th Eur. Solid-State Device Research Conf.*, pp. 600-603, ISBN 2-86332-221-4, Stuttgart, September 1997

Zappa, F., Ghioni, M., Cova, S., Samori, C. & Giudice, A. (2000). An integrated active quenching circuit for single-photon avalanche diodes. *IEEE Transactions on instrumentation and measurement*, Vol. 49, No. 6 (December 2000), pp. 1167-1175, ISSN 0018-9456

Zappa, F., Giudice, A., Ghioni, M. & Cova, S. (2002). Fully-Integrated Active-Quenching Circuit for Single-Photon Detection. *Proceedings of the 28th European Solid State Circuit Conference – ESSCIRC'02*, ISBN 88-900847-9-0, pp. 355-358, Firenze, September 2002

Zappa, F., Tisa, S., Gulinatti, A., Gallivanoni, A. & Cova, S. (2004). Monolithic CMOS detector module for photon counting and picosecond timing. *Proc. of the 34th European Solid-State Device Research Conference, ESSDERC*, ISBN 0-7803-8478, pp. 341-344, Leuven, November 2004

Zappa, F., Tisa, S., Gulinatti, A., Gallivanoni, A. & Cova, S. (2005). Complete single-photon counting and timing module in a microchip. *Optics Letters*, Vol. 30, No. 11 (June 2005), pp. 1327-1329, ISSN 0146-9592

Zappa, Z., Tosi, A., Dalla Mora A. & Tisa, S. (2009). SPICE modeling of single-photon avalanche diodes. *Sensors and Actuators A*, Vol 153, No. 2 (August 2009), pp. 197-204, ISSN 0924-4247

9

Broadband Photodetectors Based on c-Axis Tilted Layered Cobalt Oxide Thin Films

Shufang Wang* and Guangsheng Fu

Hebei Key Lab of Optic-Electronic Information and Materials, Hebei University, Baoding PR China

1. Introduction

Laser-induced voltage (LIV) effects in c-axis tilted thin films of $YBa_2Cu_3O_{7-\delta}$ (YBCO) and $La_{1-x}Ca_xMnO_3$ (LCMO) have been extensively studied in the past decades due to their potential applications in photodetectors [1-9]. Compared to the commonly used photodiodes-based photodetectors, this new type of photodetectors has the advantage of broad spectrum response ranging from ultra-violet (UV) to infrared (IR). Other advantages of this type of detectors consist in that they can be operated without cryogenic cooling and bias voltage.

The origin of the LIV signal was explained as result of the transverse thermoelectric effect which becomes effective when the c axis of YBCO or LCMO films are tilted by an angle α with respect to the film surface normal [8]. The induced voltage can be quantitatively described by the equation

$$U = \frac{l}{2d}\sin(2\alpha)\Delta S \Delta T \qquad (1)$$

Where ΔT is the temperature difference between the top and bottom of the film, which is generated by heating the film surface due to the absorption of the incident laser radiation; $\Delta S = S_{ab} - S_c$ is the difference of the seebeck coefficient in the ab-plane and along the c-axis of the film; α is the tilted angle of the film with respect to the surface normal; d is the thickness of the tilted film and ι is the laser spot diameter [1]. According to this equation, searching for materials having large anisotropy in seebeck coefficient, that is large ΔS, is a key factor for developing this new type photodetectors.

Recently, layered cobalt oxides including Na_xCoO_2, $Ca_3Co_4O_9$, $Bi_2Sr_2Co_2O_y$ and etc. have attracted great attention as promising thermoelectric materials due to their good thermoelectric performance as well as the good thermal stability, lack of sensitivity to the air and non-toxicity [10-14]. The crystal structure of these cobalt oxides consists of the conducting CoO_2 layer and the insulating Na, Ca_2CoO_3 or $Bi_2Sr_2O_4$ layer, which are alternately stacked along the c-axis. This layered structure results in a large anisotropy of the

---

* Corresponding Author

seebeck coefficient in the ab-plane and along the c-axis of the film, realizing $\Delta S = S_{ab} - S_c$ of tens of $\mu V/K$, which is about several times larger than that of YBCO ($\sim$ 10 $\mu V/K$) and hundreds times larger than that of LCMO ($\sim$ 0.22 $\mu V/K$) [3, 4]. This fact indicates the layered cobalt oxides might have potential applications in the field of high-sensitive broadband photodetectors. In this chapter, we present our investigation of LIV effects in c-axis tilted cobalt oxide thin films (Na_xCoO_2, $Ca_3Co_4O_9$ and $Bi_2Sr_2Co_2O_y$) with different pulsed laser sources with wavelength ranging from UV to NIR. The open-circuit voltage signals were detected in these films when their surfaces were irradiated by the laser light. The results demonstrate that the c-axis tilted cobalt oxide thin films have great potential applications in the broadband photodetectors.

2. Sample preparation and LIV measurements

2.1 Fabrication of c-aixs tilted Na_xCoO_2, $Ca_3Co_4O_9$ and $Bi_2Sr_2Co_2O_y$ thin films

The c-axis inclined Na_xCoO_2 thin film was obtained by epitaxially growing a layer of Na_xCoO_2 (x~0.7) on a tilted Al_2O_3 (0001) single crystal substrate by topotaxially converting an epitaxial CoO film to Na_xCoO_2 with annealing in Na vapor. A CoO film was first epitaxial grown on the tilted Al_2O_3 (0001) by pulsed laser deposition. The CoO film was then sealed in an alumina crucible with $NaHCO_3$ powder and heated to 700-750°C for 60 min to form the Na_xCoO_2 film. The c-axis tilted $Ca_3Co_4O_9$ and $Bi_2Sr_2Co_2O_y$ thin films can be grown on the tilted $LaAlO_3$ (001) or Al_2O_3 (0001) substrates with the pulsed laser deposition (PLD) or chemical solution deposition (CSD) methods. The detailed PLD and CSD fabrication parameters can be found in Ref. 15-17. Transport measurements on these films reveal that they have the room temperature seebeck coefficient comparable to that of the single crystals, suggesting good quality of these films.

Fig. 1a-c presents the x-ray diffraction (XRD) θ-2θ scans of the c-axis tilted cobalt oxide thin films on 10° tilted substrates. The offset angle ω is the angle between the c-axis direction and the substrate surface-normal direction and it is set as 10° (See the inset of Fig. 1a). Apart from the substrate peak, all peaks in these patterns can be indexed to the (00l) diffractions of the corresponding layered cobalt oxides, indicating that phase-pure c-axis tilted Na_xCoO_2, $Ca_3Co_4O_9$ and $Bi_2Sr_2Co_2O_y$ thin films are obtained and the tilted angle is 10°.

2.2 LIV measurements

Fig.2 presents the schematic illustration of the LIV measurements. Two indium or gold electrodes with the diameter of $\sim$ 0.4 mm were symmetrically deposited on the film surface along the inclined direction and they were separated by 4 mm. To prevent the generation of any electric contact effect, the electrodes were always kept in the dark. A XeCl excimer pulsed laser (λ=308 nm, t_p~20 ns) and an Nd:YAG pulsed laser (λ=532 and 1064 nm, t_p~25 ps) were used as the light sources. The incident laser beam, adjusting to 2 mm in diameter using an aperture, was directed perpendicular to the film surface at the middle position between the two electrodes. The induced lateral voltage signals were recorded using a digital oscilloscope of 500 MHz bandwidth terminated into 1 MΩ (Tektronix, TDS 3052).

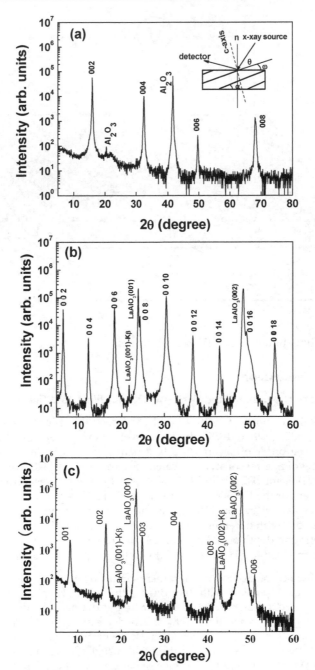

Fig. 1. XRD θ-2θ scans for the c-axis tilted (a) Na_xCoO_2, (b) $Bi_2Sr_2Co_2O_y$ and (c) $Ca_3Co_4O_9$ thin film on 10° tilted single crystal substrates. The inset of Fig. 1a is the sketch map of the XRD θ-2θ measurement and the offset angle ω is set as 10°.

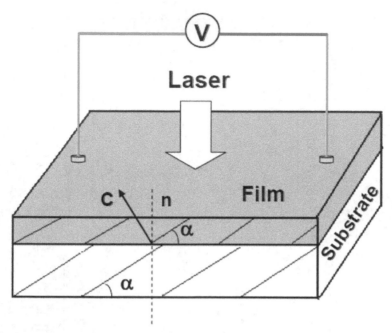

Fig. 2. Schematic illustration of the LIV measurements.

3. Results and discussion

3.1 LIV response of the c-axis tilted $Bi_2Sr_2Co_2O_y$ thin films to different laser pulses

Fig. 3a-c presents the response of a 10° tilted $Bi_2Sr_2Co_2O_y$ thin film (~ 100 nm) to the laser pulses of 308 nm, 532 nm and 1064 nm, respectively. Large open-circuit signals with the response time in the order of several hundred nanoseconds are detected for these three wavelengths. The voltage responsivity is calculated to be about 440 mV/mJ, 348 mV/mJ and 65 mV/mJ for 308 nm, 532 nm and 1064 nm pulsed radiations respectively. The stronger LIV signals obtained for the UV 308 nm and visible 532 nm lasers than that for the NIR 1064 nm laser can be explained by considering the different optical absorption and penetration depth of the radiations with different wavelength in $Bi_2Sr_2Co_2O_y$ film. The higher absorption and smaller penetration depth of 308 nm and 532 nm radiation in comparison with these of the 1064 nm radiation lead to a larger ΔT and thus a larger induced voltage.

To reduce the influence of the RC measurement circuit on the response time, a 2 Ω load resistance is connected parallel with the tilted film while keeping other experimental conditions unchanged. As shown in Fig. 4, the rise time is dramatically decreased from 100 ns shown in Fig. 2a to about 6 ns and the FWHM is also decreased from 470 ns to be about 20 ns. It should be mentioned here that the response time has a pronounced dependence on the pulse width of the incident lasers. The smaller pulse width usually leads to a faster response time. For example, the FWHM of the induced voltage signal is reduced to 1-2 ns when the same film is irradiated by the Nd:YAG picosecond laser pulse. The nanosecond-scale response of the tilted $Bi_2Sr_2Co_2O_y$ thin film to different laser pulses ranging from UV to NIR reveals that it has a potential application in broadband photodetectors with fast response.

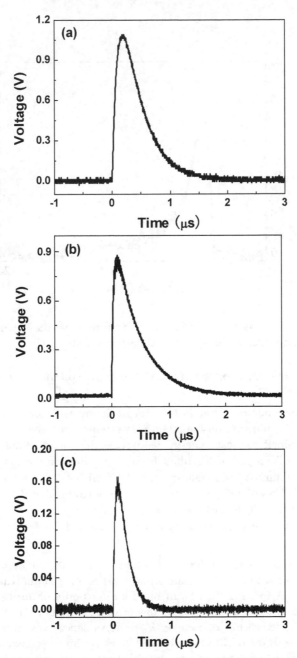

Fig. 3. LIV signals of a 10° c-axis tilted $Bi_2Sr_2Co_2O_y$ thin film (~ 100 nm) when its surface is irradiated by the pulsed lasers with different wavelength of (a) 308 nm, (b) 532 nm and (c) 1064 nm. The input impedance of an oscilloscope is 1 MΩ and the laser energy on the sample is about 2.5 mJ.

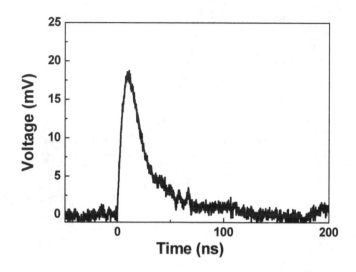

Fig. 4. LIV signal of a 10° c-axis tilted $Bi_2Sr_2Co_2O_y$ thin film under the 308 nm radiation after connecting a 2 Ω load resistance in parallel with the tilted film.

3.2 Dependence of the peak voltage of the LIV response on the laser energy, tilted angle and film thickness

Fig. 5a shows the dependence of the peak value of the open-circuit voltage signal (V_p) in the tilted $Bi_2Sr_2Co_2O_y$ thin film on the 308 nm laser energy on the film surface. A linear dependence is obtained for laser energy below the destruction limit of the film. The dependence of the LIV signal of the tilted $Bi_2Sr_2Co_2O_y$ film on the tilted angle α is also investigated and an almost linear relationship between the V_p value and α is obtained, seen in Fig. 5b. This U∝sin2α dependence again demonstrates that the LIV effect in the c-axis tilted layered cobalt oxides thin films mainly originates from the transverse thermoelectric effect since all other light-induced effects do not shown such tilt angle dependence.

Fig. 5c presents V_p as a function of the film thickness under 308 nm laser radiation. The V_P increases with the decrease of film thickness and reaches a maximum value when the film thickness is about 100 nm, and then V_p turns to a reduction with further decreasing film thickness from 100 nm to 60 nm. This is inconsistent with Eq. (1) where V_p increases monotonically with decreasing d. Similar V_p-d dependence was observed in the LIV measurements for c-axis tilted YBCO and LCMO films. An improved equation based on the plane heat source and cascade power net model was proposed to explain this abnormal behavior. Calculations based on this improved equation revealed that V_p was no more monotonic variation with the thickness d and there existed an optimum thickness corresponding to a maximum peak of the induced signal [9].

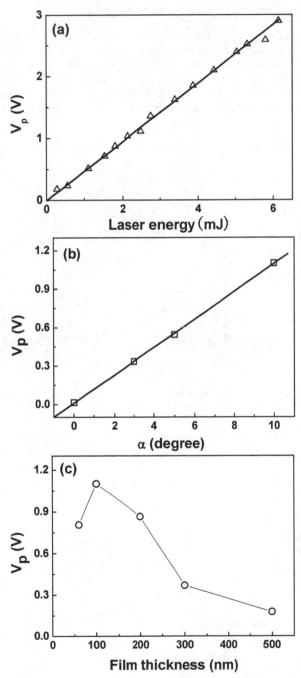

Fig. 5. Dependence of the peak value of the induced open-circuit voltage (V_p) on the (a) laser energy on the film surface, (b) tilted angle α and (c) film thickness. Solid lines are guide for eyes.

3.3 LIV signal in c-axis Ca₃Co₄O₉ and NaₓCoO₂ thin films

Similar results are also observed in the c-axis tilted $Ca_3Co_4O_9$ thin films when their surface is irradiated by the above lasers. Fig. 6 presents a typical laser-induced open-circuit voltage signal of the 10° tilted $Ca_3Co_4O_9$ thin film (~ 100 nm) under the 308 nm pulsed illumination with the laser energy on the sample of 1 mJ. Both the responsivity (~ 230 mV/mJ) and the response time (the rise time~60 ns and the FWHM~700 ns) of the LIV signal in this tilted $Ca_3Co_4O_9$ thin film is in the same order as that of the tilted $Bi_2Sr_2Co_2O_y$ thin film with the same thickness.

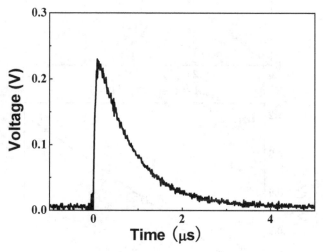

Fig. 6. A typical LIV signal of a 10° c-axis tilted $Ca_3Co_4O_9$ thin film (~ 100 nm) under the 308 nm laser radiation. The laser energy on the sample is about 1 mJ.

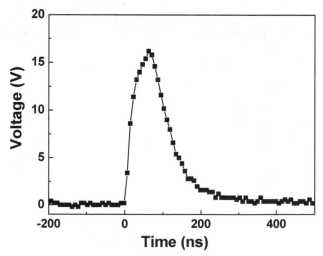

Fig. 7. A typical LIV signal of a 10° c-axis tilted Na_xCoO_2 thin film (~140 nm) under the 308 nm laser radiation. The laser energy on the sample is about 1.5 mJ.

The performed LIV measurements on the c-axis tilted Na_xCoO_2 thin films show similar dependence of V_p on the laser energy, tilted angle α and film thickness. However, the c-axis tilted Na_xCoO_2 thin films have a much larger induced voltage signal than that of the $Bi_2Sr_2Co_2O_y$ and $Ca_3Co_4O_9$ thin films. Fig. 7 illustrates the laser-induced open-circuit voltage signal of the 10° tilted Na_xCoO_2 film irradiated by the 308 nm pulsed laser with the laser energy on the sample of 1.5 mJ. A giant open-circuit voltage signal with V_p of 16.3 V is observed. The responsivity is calculated to be about 11 V/mJ. It is much larger than the responsivity of YBCO, LCMO and other two cobalt oxide thin films [1-4, 18, 19]. This might be due to the Na_xCoO_2 film has better crystal quality which corresponds to a larger ΔS, and the exact reason needs to be clarified in our future work.

4. Conclusion

In conclusion, the performed investigations of the LIV effect in c-axis tilted Na_xCoO_2, $Bi_2Sr_2Co_2O_y$ and $Ca_3Co_4O_9$ thin films at room temperature show that large open-circuit lateral voltage signals with the rise time in the order of several nanoseconds are measured when the surface of these films is irradiated by different laser sources with the wavelength ranging from UV to NIR. The obtained results suggest that c-axis tilted cobalt oxide thin films can be used for broadband photodetectors with fast response.

5. Acknowledgments

This work was partially supported by NSFC of China under Grant No. 10904030, SFC of Hebei Province under Grant No. A2009000144 and E 2006001006.

6. References

[1] H. Lengfellner, G. Kremb, A. Schnellbbgl, J. Betz, K. F. Renk, W. Prettl, Appl. Phys. Lett. 60, 601(1992)

[2] S. Zeuner, W. Prettl, H. Lengfellner, Appl. Phys. Lett. 66, 1833 (1995)

[3] Th. Zahner, R. Stierstorfer, S. Reindl, T. Schauer, A. Penzkofer, H. Lengfellner, Physica C 313, 37(1999)

[4] P.X. Zhang, W.K. Lee, G.Y. Zhang, Appl. Phys. Lett. 81, 4026(2002)

[5] H.-U. Habermeiera, X.H. Li, P.X. Zhang, B. Leibold, Sol. State Commun. 110, 473(1999)

[6] K. Zhao, M. He, G.Z. Liu, H.B. Lu, J. Phys. D: Appl. Phys. 40, 5703(2007)

[7] S. Zeuner, H. Lengfellner, W. Prettl, Phy. Rev. B 51, 11903 (1995)

[8] H. Lengfellner, S. Zeuner, W. Prettl, K. F. Renk, Europhys. Lett. 25, 375 (1994)

[9] P.X. Zhang and H.-U. Habermeier, J. Nanometers 2008, 329601 (2008)

[10] I. Terasaki, Y. Sasago, K. Uchinokura, Phys. Rev. B 56, R12685(1997)

[11] Y.Y. Wang, N.S. Rogado, R.J. Cava, N.P. Ong, Nature (London) 423, 425 (2003)

[12] M. Shikanoa, R. Funahashi, Appl. Phys. Lett. 82, 1857(2003)

[13] R. Funahashia, M. Shikano, Appl. Phys. Lett. 81, 1459(2001)

[14] K. Koumoto, I. Terasaki, R. Funahashi, MRS Bulletin 31, 206 (2006)

[15] S.F. Wang, Z.C. Zhang, L.P. He, M.J. Chen, W. Yu, G.S. Fu, Appl. Phys. Lett. 94, 162108 (2009)

[16] S.F. Wang, A. Venimadhav, S.M. Guo, K. Chen, X. X. Xi, Appl. Phys. Lett. 94, 022110 (2009)

[17] S.F. Wang, M.J. Cheng, L.P. He, W. Yu, G.S. Fu, J. Phys. D: Appl. Phys. 42, 045410 (2009)
[18] S.F. Wang, J.C. Chen, X.H. Zhao, S.Q. Zhao, L.P. He, M.J. Chen, W. Yu, J.L. Wang, G.S.
       Fu, Appl. Sur. Sci. 257, 157(2010)
[19] S.F. Wang, J.C. Chen, S.R. Zhao, L.P. He, M.J. Chen, W. Yu, J.L. Wang, G.S. Fu, Chin.
       Phys. B 19, 107201 (2010)

Permissions

The contributors of this book come from diverse backgrounds, making this book a truly international effort. This book will bring forth new frontiers with its revolutionizing research information and detailed analysis of the nascent developments around the world.

We would like to thank Dr. Sanka Gateva, for lending her expertise to make the book truly unique. She has played a crucial role in the development of this book. Without her invaluable contribution this book wouldn't have been possible. She has made vital efforts to compile up to date information on the varied aspects of this subject to make this book a valuable addition to the collection of many professionals and students.

This book was conceptualized with the vision of imparting up-to-date information and advanced data in this field. To ensure the same, a matchless editorial board was set up. Every individual on the board went through rigorous rounds of assessment to prove their worth. After which they invested a large part of their time researching and compiling the most relevant data for our readers. Conferences and sessions were held from time to time between the editorial board and the contributing authors to present the data in the most comprehensible form. The editorial team has worked tirelessly to provide valuable and valid information to help people across the globe.

Every chapter published in this book has been scrutinized by our experts. Their significance has been extensively debated. The topics covered herein carry significant findings which will fuel the growth of the discipline. They may even be implemented as practical applications or may be referred to as a beginning point for another development. Chapters in this book were first published by InTech; hereby published with permission under the Creative Commons Attribution License or equivalent.

The editorial board has been involved in producing this book since its inception. They have spent rigorous hours researching and exploring the diverse topics which have resulted in the successful publishing of this book. They have passed on their knowledge of decades through this book. To expedite this challenging task, the publisher supported the team at every step. A small team of assistant editors was also appointed to further simplify the editing procedure and attain best results for the readers.

Our editorial team has been hand-picked from every corner of the world. Their multi-ethnicity adds dynamic inputs to the discussions which result in innovative outcomes. These outcomes are then further discussed with the researchers and contributors who give their valuable feedback and opinion regarding the same. The feedback is then collaborated with the researches and they are edited in a comprehensive manner to aid the understanding of the subject.

Apart from the editorial board, the designing team has also invested a significant amount of their time in understanding the subject and creating the most relevant covers. They scrutinized every image to scout for the most suitable representation of the subject and create an appropriate cover for the book.

The publishing team has been involved in this book since its early stages. They were actively engaged in every process, be it collecting the data, connecting with the contributors or procuring relevant information. The team has been an ardent support to the editorial, designing and production team. Their endless efforts to recruit the best for this project, has resulted in the accomplishment of this book. They are a veteran in the field of academics and their pool of knowledge is as vast as their experience in printing. Their expertise and guidance has proved useful at every step. Their uncompromising quality standards have made this book an exceptional effort. Their encouragement from time to time has been an inspiration for everyone.

The publisher and the editorial board hope that this book will prove to be a valuable piece of knowledge for researchers, students, practitioners and scholars across the globe.

List of Contributors

N.N. Jandow, H. Abu Hassan, F.K. Yam and K. Ibrahim
Nano-Optoelectronics Research and Technology Laboratory School of Physics Universiti Sains Malaysia, Minden, Penang, Malaysia

Maurizio Casalino, Luigi Sirleto, Mario Iodice and Giuseppe Coppola
Istituto per la Microelettronica e Microsistemi, Consiglio Nazionale delle Ricerche, Naples, Italy

J.A. Luna-López, J. Carrillo-López, F. Flores-Gracia and D.E. Vázquez Valerdi
Science Institute-Research Center for Semiconductor Devices-Autonomous, Benemérita University of Puebla, México

M. Aceves-Mijares
Department of Electronics, National Institute of Astrophysics, Optics and Electronics INAOE, México

A. Morales-Sánchez
Centro de Investigación en Materiales Avanzados S. C., Unidad Monterrey-PIIT, Apodaca, Nuevo León, México

Yu Zhu Gao
College of Electronics and Information Engineering, Tongji University, Shanghai, China

M. Ghanashyam Krishna and Surya P. Tewari
Advanced Centre of Research in High Energy Materials, India
School of Physics, University of Hyderabad, Hyderabad, India

Sachin D. Kshirsagar
Advanced Centre of Research in High Energy Materials, India

Marco Ramilli, Alessia Allevi, LucaNardo and Massimo Caccia
Dipartimento di Fisica e Matematica - Università degli Studi dell'Insubria, Italy

Maria Bondani
Istituto di Fotonica e Nanotecnologie - Consiglio Nazionale delle Ricerche, Italy

Takeji Ueda and Susumu Komiyama
The University of Tokyo, Japan

Anna Vilà, Anna Arbat, Eva Vilella and Angel Dieguez
Electronics Department, University of Barcelona, Spain

Shufang Wang and Guangsheng Fu
Hebei Key Lab of Optic-Electronic Information and Materials, Hebei University, Baoding, PR China